For Mark and Kelly

Table of Contents
Dr. Phil's RF Electronics
©2016

CHAPTER 1: THE RADIO BUILDING HOBBY

CHAPTER 2: SINGLE TRANSISTOR RADIOS

CHAPTER 3: SINGLE VACUUM TUBE RADIOS

This page intentionally blank.

1.1 Radio Building Components
THE RADIO BUILDING HOBBY
©2015

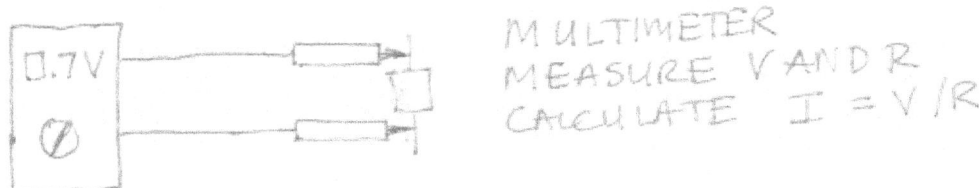

Even simple radio circuits are complex and to fully appreciate might require advanced mathematics such as: calculus, Fourier analysis, and integral transformations. That said, it is much easier to solder up a circuit and take multimeter measurements. Measuring the voltage (V) and the resistance (R) of an electrical component, the current (I) through it can be simply calculated.

1. THE ANTENNA

Radio frequencies (RF), or electromagnetic waves, enters a circuit via an antenna. The antenna can be 50 feet of wire. On mediumwave (MW), a loop antenna offers directionality: it can be pointed at a station. On shortwave (SW), a loop antenna can exclude local noise. On FM, dipoles are used. **The impedance of free space is ~377 ohms**. Free space's low impedance matches with the input of a common-base, common-gate, or common-grid amplifier. A small variable capacitor (trimmer) is often used to introduce radio energy into the tank circuit. This prevents loading the tank. Another type of input is inductive coupling. By using fewer antenna turns than tank turns, voltage is increased (current is decreased). This works well with voltage-controlled amplification devices such as a JFET or vacuum tube. A BJT is a current-controlled amplifier.

2. THE TANK

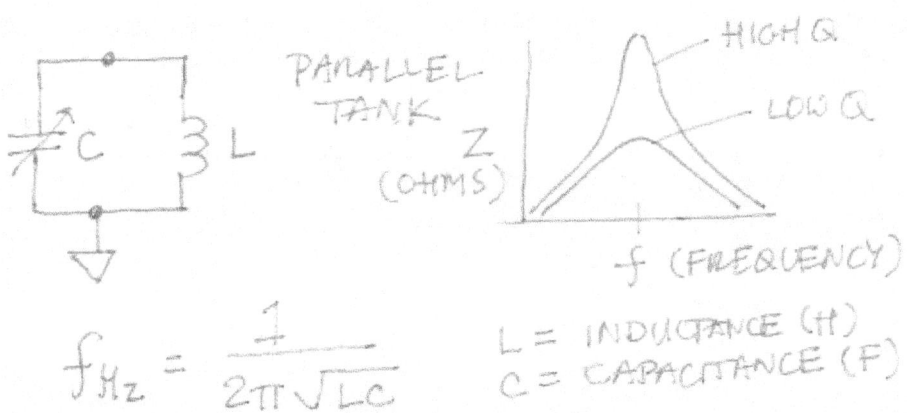

$$f_{Hz} = \frac{1}{2\pi\sqrt{LC}}$$

L = INDUCTANCE (H)
C = CAPACITANCE (F)

On simple radios, **selectivity come from the Q of the tank.** A Q of 100 is easy to achieve. A Q of 1000 is more difficult. Using regeneration (positive feedback), a low Q tank can achieve a high Q. This is called Q-multiplication. Selectivity allows hearing the station of interest, alone. A parallel, inductive-capacitative (LC) tank allows the passage of many frequencies, centered around the tuned frequency. An extremely high Q can be disadvantageous because the circuit will oscillate and only tune one frequency. Q can be spoiled or reduced by placing resistance across the tank. Selectivity using a tank becomes harder as the tuned frequency increases: ex. SW. Crystal sets often tune MW. A tank can use a 365 pF variable tuning capacitor. On medium wave (520 kHz to 1710 kHz), the inductor needs to be about 256 µH. On shortwave (3 MHz to 30 MHz), the inductor needs to be about 7.7 µH. On FM, small values of inductance and capacitance are needed. For example, a 20 pF variable capacitor and 163.5 nH. **The tank dumps most radio energy to ground except the frequency of interest and its surrounding voice frequencies.** The data that represents human voice spans from about 300 to 3400 Hz. The equation above shows that the tank's frequency is related to its inductance (L) and capacitance (C) values. Maximum resistance occurs at resonance. Below is a series LC tank with low resistance.

$$Z = \sqrt{R^2 + (X_L - X_C)^2}$$

(OHMS)

SERIES LC

3. DETECTION DIODE: 1N34A

ANODE   CATHODE

IN ——— 0.0 V

OUT ——— 0.3 V

NO REVERSE FLOW

1N34A

0.3 V Ge

"VOLTAGE DROP" FORWARD

AM stations with a carrier and two sidebands can be demodulated using a diode or other non-linear device. Beyond the detector, there must be a path for radio energy to reach ground. This can be a tiny radio frequency capacitor. **The 1N34A, a germanium point contact diode, is a common detector.** The 1N34A is classified as giving "an efficient and excellent linearity when used in TV image detection, FM detection, and radio AM detection." Maximum diode reverse voltage is 20 Volts and maximum forward current is 150 mA. An active device, such as a transistor or vacuum tube may also be used for detection. Detection turns radio waves (radio frequencies, RF) into audio waves (audio frequencies, audio, AF). A diode can act as a clamp (cutting signals over 0.3 or 0.7 Volts) or clipper (resistor, diode with pot under, to clip positive or negative peaks).

IN   OUT

IN   RADIO

OUT   AUDIO

4. RESISTORS

—WW— RESISTOR $V = I \cdot R$

Ohm's law : VOLTAGE = Current × Resistance
              (VOLTS)   (AMPS)   (OHMS)

Resistors reduce current or electrical flow. The value of many components in a radio circuit do not need to be exact: a range of values will work. It is important to look at both capacitors and inductors as resistors: via their capacitative reactance and inductive reactance, that varies with frequency. Unlike capacitors and inductors, a resistor's value does not change with frequency. And a resistor works on direct current (DC). DC is the type of current from a battery. Whereas, radio and audio are forms of alternating current (AC): although they may ride on DC. Variable resistors are called potentiometers or pots. Resistors in series add; resulting in increased resistance. Resistors in parallel result in lowered resistance. The math is simple but a multimeter will show the value in seconds. Ohm's law states that voltage (E, in Volts) equals current (I, in Amperes) times resistance (R, in Ohms). The largest power dump is into the part in series with the most resistance. A voltage in mV equals current in mA times resistance. Electrons flow from ground to the positive supply; however, "conventional current" flows the opposite (from plus power to ground). Power (P) equals Volts (V) times Amps (I). A hertz (Hz) is one cycle per second.

RESISTORS ADD IN SERIES   —WW— VARIABLE
                                  RESISTOR

—WW——WW—• $R = R1 + R2$
  R1   R2

5. SMALL CAPACITORS

CAPACITORS ADD IN PARALLEL.

C1 ǂ ǂ C2 $C = C1 + C2$ VARIABLE CAPACITOR

Capacitors store electrons. Small capacitors have values of 1000 pF (102k or 0.001 µF) to 10,000 pF (103k or 0.010 µF), or less. **Small capacitors block both direct current and audio frequencies**. Small capacitors block DC and audio. Using 0 Hz, or DC, in an equation for capacitative reactance gives a value of infinity ohms. Using a value of 1000 pF, the capacitative reactance at 300 Hz is 530k ohms and at 3400 Hz is 46k ohms. These are high resistances. However, at 1.115 MHz (the middle of the MW band), a 1000 pF capacitor is seen as only 142 ohms. At 16.5 MHz (the middle of the SW band), a 1000 pF capacitor is seen as 9.6 ohms. And at 98 MHz (the middle of the FM band), a 1000 pF capacitor is seen as only 1.6 ohms. **Small capacitors are used to pass radio frequencies**. Capacitors in parallel add; resulting in increased capacitance. Capacitors in series result in lowered capacitance. They can be measured.

IN —||—•→ OUT   IN —WW—•→ OUT
       HIGH-PASS   LOW-PASS
       FILTER   FILTER

6. LARGE CAPACITORS

$$X_C = \frac{1}{2\pi f C} \; (ohms)$$

$\pi = 3.1415$

f = frequency

C = in Farad

CAPACITOR

REGULAR (SMALL)

ELECTROLYTIC (LARGE)

Large capacitors have values from 0.22 µF to 1000 µF, or higher. In the range of 5 to 1000 uF, the capacitor is an electrolytic type. **Large capacitors will block direct current but still carry audio.** Electrolytic capacitors have plus and minus poles and must be connected properly. The minus end may be soldered to ground; while, the plus end may be soldered to the positive power supply. This helps a battery keep up with high, transient, current demands. Large, non-electrolytic capacitors (ex. 0.22 uF ceramic discs) can be used to pass both audio and radio frequencies. **Do not trust an electrolytic capacitor to pass radio frequencies.** Use a small capacitor (0.10 µF) in parallel to pass radio. At DC, capacitative reactance is infinity ohms. Using a 100 uF capacitor, an audio frequency of 300 Hz sees 5.3 ohms; and 3400 Hz sees 0.5 ohms. Audio passes in large capacitors. Small (RF) or large (audio) capacitors "couple" active devices: they remove DC and just let the signal (RF or audio) through. X_C is capacitative reactance: the equation is above. X_L is inductive reactance: the equation is below. Both give resistance values in ohms.

7. SMALL INDUCTORS

INDUCTORS ADD IN SERIES

$L = L1 + L2$

L1   L2

TOROID "L"

$$X_L = 2\pi f L \; (ohms)$$

$\pi = 3.1415$ f = frequency L = in Henry

Inductors induce an electromagnetic field. They resist AC. A small inductor is typically used in a tank circuit. **Small inductors easily pass both direct current and audio frequencies.** The 256 uH tank inductor used on MW is seen as 1.79k ohms by 1115 kHz radio waves. This same tank is seen as 0 ohms at DC, 0.5 ohms at 300 Hz, and 5.5 ohms at 3400 Hz. Small inductors can be created using toroids (see above) or via turns of wire in the air. Inductors in series add; resulting in increased inductance. Inductors in parallel will result in lowered inductance.

8. MEDIUM INDUCTORS

Medium inductors can be used to block or choke radio frequencies. A typical MW choke has a value of 2.5 mH. A 1115 kHz signal sees 2.5 mH as 17.5k ohms. However, DC sees this same 2.5 mH as almost 0 ohms. While 300 Hz and 3400 Hz see it as 4.7 ohms and 53 ohms, respectively. On SW, at 16.5 MHz, a choke of 2.5 mH is seen as 259k ohms! Often smaller, 1 mH chokes are used on SW. On FM, at 98 MHz, a 2.5 mH choke is seen as 1.5M ohms. On FM, a few turns of wire (formed by winding them on a pencil) can be used as a radio frequency choke. **Medium inductors pass DC and audio.** RF chokes can be made by using ferrite toroids.

9. LARGE INDUCTORS

Large inductors are used as audio chokes. Most designers avoid using an audio choke. These chokes are large and expensive. A perfect large inductors should block radio frequencies. However, in practice, a dedicated radio frequency choke should be used. Large inductors can have a fair amount of resistance to direct current, for example: 1500 ohms. This is due to them using a very long and thin wire. The 24H (Henry) T-725 transformer can be used as an audio choke. This 24H is seen as 45k ohms at 300 Hz and 512k ohms at 3400 Hz. These are high resistances. Think of a high resistance as being like air, it is an "open" circuit; meaning, it is as if it is not connected.

10. INDUCTOR PROBLEMS

A friend of mine stated: "**Even a piece of wire has capacitance and inductance.**" In an ideal simple radio, the only inductor used would be in the tank. The reason is that wire wound together can have capacitance. This capacitance can pass radio. This is how a large inductor (audio choke or audio transformer) can end up leaking radio frequencies. And why dedicated radio frequency chokes are used. An even better solution is to use multiple resistive-capacitative (RC) low-pass filters to choke radio frequencies. Two RC filters, in series, can achieve about -65 dB.

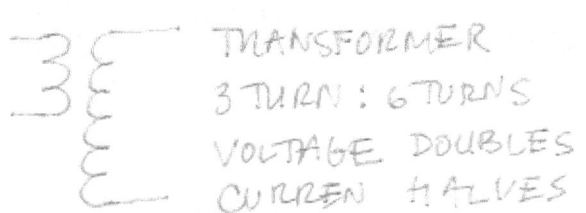

11. RF TRANSFORMERS

RF transformers can boost voltage. An inductively coupled antenna input or wire turns into a tank, is a RF transformer. Voltage is increased (and current decreased) when there are less turns from the antenna than in the tank. Capacitative coupling, using a small trimmer cap, is often desirable because a radio may need multiple inductive taps for the antenna to be effective over the entire tuning range. A common toroid for MW transformers is "61" material (no color code); for SW is "2" material (red color code); and for FM is "10" material (gray color code). Transformers can only transfer AC. Any direct current simply flows through the wire making up the transformer's primary coil. DC is not induced in the secondary winding of an RF or audio transformer or speaker.

12. AUDIO TRANSFORMERS

The Bogen T-725 is a valuable audio transformer. A T-725 is a 24H transformer with 9 taps and is capable of transforming an 8 ohm earphone to 45k ohms at 300 Hz and 512k ohms at 3400 Hz. Double these for 16 ohm earphones. Direct current (DC) r1000esistance is ~1420 ohms. Borden sells a 1k:50k audio transformer with 1200 ohms of primary resistance (DC). The purpose of an audio transformer is to increase the tiny 8 or 16 ohm load of an earphone to 45k or 90k ohms. This is done because voltage gain in active devices is proportional to their load resistance.

color	DC ohms	Xl at 900 Hz
white	1395	135.7k
gray	867	67.9k
violet	505	33.9k
blue	257	17.0k
green	84	8.9k
yellow	57	4.4k
orange	39	2.3k
red	26	1.1k
brown	18	565

13. AUDIO

Ｏ HEADPHONE   ┴ 16nF = CRYSTAL EAR
Ｏ SYMBOL   ┬

In general, sound powered phones are no longer needed. Which is good because sound powered phones are costly and often require being rebuilt. They are available on Ebay. The Bogen T-725 transformer mated to sensitive ear-buds (ex. the 16-ohm Koss Sparkplug, 112dB SPL/1mW) are sufficient for simple radios. Crystal earphones are sensitive but sound tinny (high pitched). Crystal earphones can be viewed, electrically, as a small (~16 nF) capacitor. DC cannot flow through them, RF can. Using piezo elements (ex. KBT-44SB-1A) as earphones is not recommended. They are sensitive but can result in hearing loss due to very loud "clicks". Another audio option is to buy or make an LM386 (more gain) or TDA7052 (less parts) amplified speaker. This allows testing a simple radio or crystal set without straining to hear. NOS Chinese military headsets (4400-ohm), magnetic earphones (2000-ohm), and Telex magnetic headsets (600-ohm) are available at Borden Radio Company. An amplified speaker can be built out of an inexpensive Sony S10MK2 AM/FM radio by feeding the signal into the audio amplifier pin on the CXA1019S chip. Note that the ground lines (battery minuses) of the circuit and the radio must be attached.

14. POWER SOURCES

┴ ─   9V   500 mAh
 ─ BATTERY 1.5V "D" 13000 mAh ┬
┬ ＋   ▽ GROUND

Wall wort power supplies can induce noise and should be avoided. **Use batteries.** A JFET typically requires 9 Volts. A BJT can work on 1.5 Volts. A pencil vacuum tube can run on 18 to 27 Volts (2 or 3 nine Volt batteries) of plate. Batteries in series add voltages. The heater can run on a 1.5 Volt D-cell. Two 1.5 Volt "D" batteries, in series, will power a 3GK5 for 30 hours. This tube is the same as a 6GK5 but with a 3 Volt heater that draws 450 mA. This is a safe way to experiment with a high-performance tube. Do not use over three 9 Volt batteries, in series, for plate voltage. Avoid dangerous voltages and wall wort power supplies. Tubes will work at low voltages: never apply hundreds of volts to tubes, it is not needed. **Safety first.** Note that some schematics count on their being bypass capacitors in the power supply. These capacitors send audio and radio frequencies to ground. Using batteries, bypass capacitors may have to be added back to the designs. **Always wear polycarbonate eye protection when building radios.** If +9V is attached to ground (it is now at "ground"), the battery "-" is now at minus 9V (voltage is relative).

15. BUILDING TANK INDUCTORS

It is easy to get started at building inductors. Order from Amidon a spool #24 AWG enamel coated magnet wire and a few T106-2 red toroids for shortwave. Cut a few feet of wire and start wrapping. Leave enough wire that it is possible to increase inductance. Scrape the enamel off the ends using a tiny pocket knife, heat for several seconds, and solder tin them. There is plenty of wire to practice. On MW it is better to build a spider coil using Litz wire (lookup skin effect). These inductors take more skill. Shortwave regenerative radios are exciting to build because the stations heard can be anywhere in the world. MW travels best at night. SW has both night bands (5.7 to 10.0 MHz) and day bands (11.5 to 17.9 MHz). FM is best heard with a large vertical antenna or a dipole. FM is line of sight. The wavelength in meters (3.28 feet) is 300 divided by the frequency in MHz. This means that a MW station at 1 MHz has a 300 meter wavelength or it is about 3 football fields in length. A SW station at 10 MHz has a 30 meter wavelength or about 100 feet in length.

16. PARTS BIN COMPONENTS

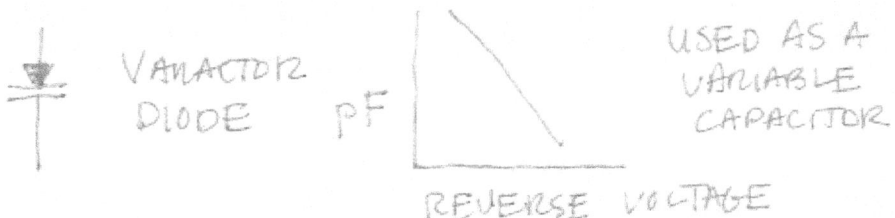

The following is a list of items used to build radios: eye protection, a soldering iron, iron stand and sponge, desoldering braid and bulb, and solder. Wire, buss wire, clippers, and needle-nose pliers. Basic electrical components: wire, resistors, capacitors, electrolytic capacitors, varactors (see drawing above), variable resistors (potentiometers), RF chokes, half-inch socket for potentiometer mounting, variable capacitors (365 pF and trimmers), D and 9 Volt batteries, and battery holders. Toroid building items: coated wire of #20 AWG, #24 AWG, and #26 AWG. Shortwave iron-powder toroids: T16-2, T25-2, T37-2, T50-2, T80-2, T106-2, T130-2 (type "2" material, red). FM iron-powder toroids: T25-10, T37-10, T50-10, T80-10 (type "10" material, gray). MW ferrite toroids: FT-23-77, FT-37-77, FT-50-77, FT-82-77 (type "77" material). SW chokes: FT-82-75. Others MW toroids: FT-82-43, FT-82-61, FT-114-61. Active devices (see list below): transistors, tubes, and tube sockets. Audio transformers (Bogen T725), Koss Sparkplug 16-ohm earphones, crytsal earphones, and an amplified speaker. A multimeter, storage bins, and a spotter radio: GP5/SSB or DE1103 (AM/SSB AM/SW/FM). And last, but not least, a set of "second hands" clamps.

17. DC VERUS AC

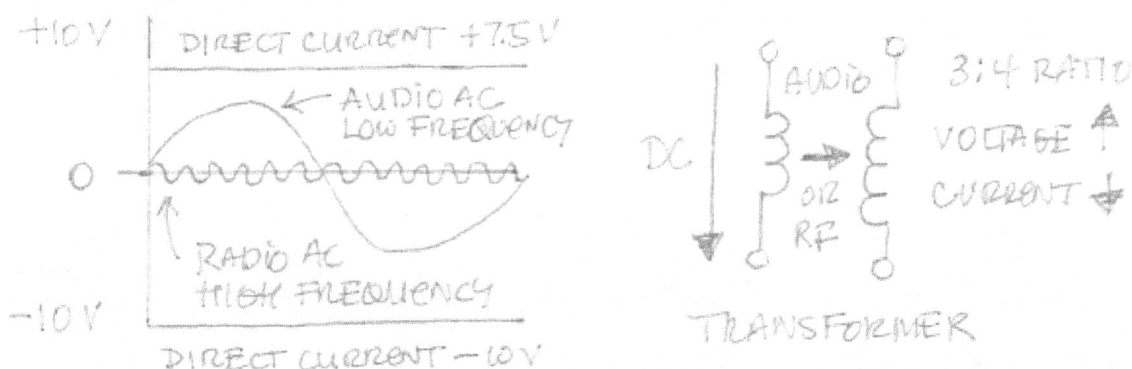

Above see direct current (DC) and alternating current: fast AC (radio) and slow AC (audio). A transformer passes DC in its primary winding: only AC is transferred to the secondary winding. A larger to smaller larger turns ratio will drop voltage but increase current. The opposite is also true.

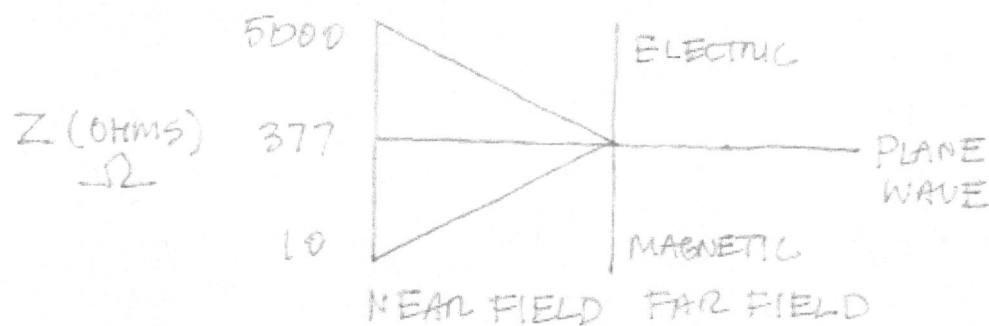

Above shows the near field with high impedance electric field waves and low impedance magnetic field waves. The far field consists of a plane wave with an impedance of 377 ohms.

11

1.2 Component Circuit Analysis
THE RADIO BUILDING HOBBY
©2015

1. OHM'S LAW

Ohm's law : VOLTAGE = Current × Resistance
             (VOLTS)   (AMPS)   (OHMS)

Ohm's law states that the current in a component (in amps) multiplied by the resistance of a component (in ohms) is equal to the voltage across a component (in Volts). See above equation.

2. POWER DUMPING

POWER = VOLTS × AMPS

$$\frac{10V}{1\Omega+9\Omega} = 1A$$

$1\Omega \times 1A = 1V$
$9\Omega \times 1A = 9V$
$1A \times 1V = 1 Watt$
$1A \times 9V = 9 watt$

$10V \div 1\Omega = 10A$
$10V \div 9\Omega = 1.1A$
CURRENT

$10V \times 10A = 100W$
$10V \times 1.1A = 11W$
POWER

ELECTRON FLOW

Electrons flow from ground to the positive supply. Via Ohm's law, in series, the component with a high resistance gets the most energy being dumped into it. In parallel, the smallest resistance gets the most energy dumped into it. This is important in circuits analysis as often one component gets most of the energy and dominates how the circuit works. Example are above.

3. KIRCHHOFF'S VOLTAGE AND CURRENT LAWS

KIRCHHOFF'S VOLTAGE LAW

$$V_{SUPPLY} = V_{R1} + V_{R2} + V_{R3}$$

Kirchhoff's voltage law (KVL) states that: the sum of the voltages (V) around a loop is equal to zero. The voltage in the supply above is equal to the sum of the voltage drops across resistors R1, R2, and R3. Kirchhoff's current law (KCL) states that: the sum of the currents (I) flowing into a point equals the sum of the currents flowing out of a point. The current entering the point (I1 plus I2) must equal the current exiting the point (I3 plus I4). The point is shown, below, as a circle.

KIRCHHOFF'S CURRENT LAW

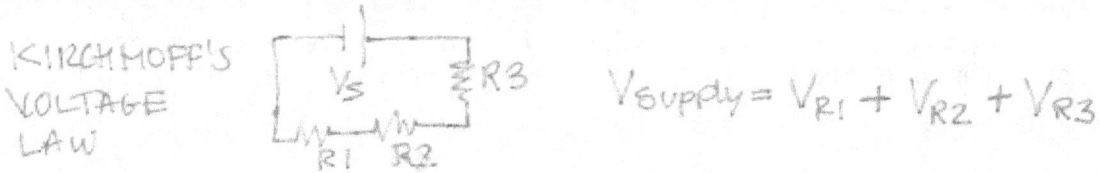

$$I_1 + I_2 = I_3 + I_4$$

4. RADIO FREQUENCY (RF) CIRCUIT ANALYSIS

Each circuit should be analyzed for RF, DC, and audio. Radio is AC or alternating current. When analyzing the RF part of each circuit the antenna remains. Ideally, the parallel LC tank dumps all RF to ground except the frequency tuned, and its surrounding frequencies. For the frequency tuned the tank is as if it is not there (high Z). A diode turns RF into audio. But it also does something not often talked about. The other half of the RF is trapped above the diode (plus 0.3V of the passed side). This can be used as a "perfect" choke. Resistors remain. Small capacitors turns into a piece of wire. Large capacitors turn into a piece of wire. Each small inductor will turn into a resistor. Medium inductors are RF chokes: high resistance. Each large inductor will turn into an open circuit (not connected). An RF transformer will pass RF into the other winding. Turns ratios can be used to increase voltage or current (at the other's expense). An audio transformer will also turn into an open circuit (not connected). It is not good to rely on large capacitors to dump RF to ground. Nor is it good to rely on large transformers to block RF. Use a small capacitor or an RF choke, respectively. Circuits must filter, amplify, and detect RF. The filter can be a tank, the amplifier can be a transistor or tube, and the detector can be the same active device or a diode.

5. AUDIO FREQUENCY (AF) CIRCUIT ANALYSIS

Each circuit should be analyzed for audio, DC, RF. Audio frequencies are AC or alternating current. When analyzing the audio part of each circuit the antenna can be erased. A parallel LC tank turns into a wire via its small inductor. A diode can actually pass audio that is already present. Resistors remain. Small capacitors turns into an open circuit (not connected), due to large resistance. Large capacitors turn into a piece of wire. Each small inductor will turn into a piece of wire. Each large inductor will turn into a resistor. Very large inductors will choke audio. One winding of the Bogen transformer is 24H of inductance and, essentially, an audio choke. A resistor load can be placed across the other winding to boost resistance. This is how 8-ohm phones are boosted to 1,000 ohms by small audio transformers. This is done because voltage gain (and hence power gain) is proportional to the active device's load resistance. Each RF transformer will also turn into a wire. Audio has no effect in the other winding. And an audio transformer will transfer audio from one side of the winding to the other. Typically with a tremendous increase in current. If audio is riding on a DC current, this DC current cannot: 1) pass through the capacitance of a crystal earphone, and 2) cannot be transferred to the other side of an RF transformer. The DC component is, essentially, filtered out by not being able to pass to the secondary winding.

6. DIRECT CURRENT (DC) CIRCUIT ANALYSIS

Each circuit should be analyzed for direct current (DC), radio frequencies (RF), and audio frequencies (audio). When analyzing the DC part of each circuit the antenna can be erased. The parallel LC tank turns into a wire via its small inductor. A diode can only flow DC in one direction (plus to cathode or arrow symbol; minus to anode or line symbol). There is a voltage drop of 0.3 Volts for germanium (1N34A) and 0.7 Volts for silicon diodes. Resistors remain. Small capacitors turns into an open circuit (not connected). Large capacitors turn into an open circuit. This will eliminate many lines on the schematic. Each small inductor will turn into a piece of wire. Each large inductor will turn into a resistor. An RF transformer will turn into a piece of wire. DC cannot transfer in a transformer. Each audio transformer will turn into a resistor. This is because DC cannot cross or transform itself into the other coil winding or have any effect on the earphones.

As a side note: Capacitors store energy in an electric (electrostatic) field. Current leads a capacitors voltage by 90 degrees. Inductors store energy in a magnetic field. Capacitors with straight-line capacitance look like half-circles. Capacitors with straight-line frequency look like a bird's wing. Current lags an inductor's voltage by 90 degrees. An LC (inductor-capacitor) tank can be thought of as a pendulum.

7. COMPONENT IMPEDANCE OVERVIEW

component	example	DC	audio	radio
antenna	wire	open	open	incoming
tank	parallel	pass	pass	tunes
diode	1N34A	pass	pass	detects
resistor	any	same	same	same
small cap	.01 uF	block	block	pass
large cap	10 uF	block	pass	pass
small coil	4 uH	pass	pass	tank
medium coil	1 mH	pass	pass	block
large coil	24 H	pass	block	block
RF transformer	tickler	pass	pass	transfer
audio transformer	T-725	pass	transfer	block

component	value	ohms DC	ohms audio 300 Hz	ohms audio 3000 Hz
resistor	10k ohm	10k	10k	10k
tiny cap	100 pF	infinite	5.3M	530k
small cap	.01 uF	infinite	53k	5.3k
large cap	10 uF	infinite	53	5.3
small coil	4 uH	0	0	0
medium coil	1 mH	0	1.9	19
large coil	24 H	1420	45k	450k

component	value	ohms 1MHz RF	ohms 16 MHz RF	ohms 98 MHz RF
resistor	10k ohm	10k	10k	10k
tiny cap	100 pF	1.6k	99	16
small cap	.01 uF	15	1	0
large cap	10 uF	0	0	0
small coil	4 uH	25	402	2.5k
medium coil	1 mH	6.3k	100k	610k
large coil	24 H	150M	2.4G	14G

Study the three charts above. The first chart is an overview; the other two have exact (ohm) values. The second chart shows resistances to DC and low and high audio frequencies. The third chart shows resistances to MW, SW, and FM radio frequencies. This is how a radio circuit sees components. The radio circuit will contain: direct current (DC), radio frequency energy (RF), and audio frequencies (AF). And each of these sees each component a different way. A high number of ohms will make that part of the circuit look like it is not connected. In series, a low number of ohms will mean that not much energy is dumped into that component. In parallel, a high number of ohms will mean that not much energy is dumped into that component. The term Z or impedance is in ohms and can be thought of as a summary resistance made up of purely resistive, capacitative reactance, and inductive reactance components. Note that DC can only pass a diode in one direction: positive to anode (triangle) and negative to cathode (line). The value 45k at 300 Hz for 24H shows good transfer (to the headphones or earbuds) of the audio in a transformer.

1.3 Receiver Topology
THE RADIO BUILDING HOBBY
©2015

1. CRYSTAL RADIO

A simple crystal set may consist of an antenna, tank, second tank, diode detector, followed by a capacitor to ground, resistor to ground, and crystal earphones to ground. A crystal set does not contain an active (gain providing) device. Voltage gain is achieved via transformer turns ratios; but at the expense of lowered current. High performance crystal sets take skill to create and the learning curve is steep. Dual tuned tanks, inductively coupled, are used for added selectivity. The arrow in the schematic shows that the tanks can be moved closer or further apart (coupling).

2. REFLEX RADIO

A reflex circuit uses a single active device to provide both radio and audio frequency amplification. Radio is sent to an amplifier, detected, and then audio is fed back into the same amplifier. Reflex sets work well on MW with high Q tanks for selectivity. Selectivity is not ideal on SW. Reflex circuits have stable amplification, as it splits amplification into radio and audio gain. However, selectivity is only as good as the Q of the tank. The simplified example above uses an RF choke (RFC) to pass audio but block radio. A small capacitor passes radio but blocks audio.

3. DIRECT CONVERSION RADIO

Direct conversion mixes a signal directly down to audio. For example, a station carrier at 910 kHz is mixed with 910 kHz and the result is audio. This is then fed into a powerful audio amplifier. Direct conversion is also called zero-IF, autodyne, or homodyne. The primary problem with direct conversion is that the radio needs a lot of audio gain. It is not ideal for "AM" reception.

4. SUPERHETERODYNE RADIO (SUPERHET)

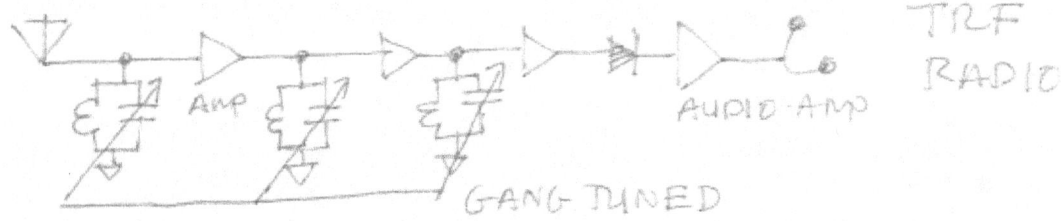

Superheterodyning uses a mixer to mix radio frequencies (RF) with a local oscillator (LO). This creates both sum and difference frequencies. This mixes the signal down to a set frequency, where an intermediate frequency (IF) filter (ex. 455 kHz) and IF amplifier can be employed for good selectivity and gain. For simple radios, the lack of a radio frequency amplifier is not a huge setback: long wires can be used. However, single active device superhets, using a pentagrid converter vacuum tube, have only conversion gain (unless reflexed). Amplification is typically achieved by using vacuum tubes at lethal plate voltages. A single tube superhet is not the greatest DX (distance reception) radio. It is also a challenge to build due to RF tuning having to track the tuning of the local oscillator. In down-conversion a signal at 1.500 MHz is heard by mixing it with 1.045 MHz, resulting in both 2.545 MHz and 0.455 MHz. The second frequency, 455 kHz, passes through a 455-kHz IF filter. The filter may have a bandwidth of 6 kHz (AM) or 3 kHz (SSB). Both signals are amplitude modulated. AM here refers to a carrier with two (audio data) sidebands.

5. DOUBLE-CONVERSION SUPERHET RADIO

Modern communications receivers (and quality portables) use an RF amplifier before their first up-converting mixer. The local oscillator has a variable frequency. RF is filtered, to remove the image, and then amplified. This feeds a second down-converting mixer (at a fixed frequency) and IF filter. This filter is used for selectivity; because tight filters are easier to make at lower frequencies (why 455-kHz is often used). The RF is amplified and sent to a detector (diode for AM or mixer/BFO for SSB) and then enters an audio amplification stage (often an integrated circuit).

6. TUNED RADIO FREQUENCY (TRF) RADIO

A TRF or tuned radio frequency radio sends the RF through multiple stages of tanks separated by RF amplifiers. There are typically three tuned stages. All three tanks are tuned simultaneously via ganged variable capacitors. The resulting amplified RF is fed into a detector (diode) and then an audio amplifier. A TRF radio works best on MW and is less effective on SW. For a tank to get the same bandwidth at 10 MHz takes 10 times the Q factor than it takes at 1 MHz.

7. REGENERATIVE RADIO

The regenerative receiver is the bread and butter of single active device radios. Regeneration is positive feedback. This positive feedback results in both amplification and selectivity. These are the two critical things that a radio must possess. A regenerative set is basically an oscillator set just before oscillation. Oscillators will be covered, in depth, below. An antenna feeds a tank circuit and RF is sent into an amplifier. The output of this amplifier is fed back into the tank via inductance (turns of wire), capacitance, or resistance (atypical). This positive feedback or regeneration, causes up to 78 dB of gain (or 8000; 1933 QST data, 85 dB for Morse code) and high-Q selectivity. The circuit can be over-regenerated. This will cause oscillation. In this mode the regenerative radio will act as a **direct-conversion circuit**. Selectivity may be enhanced but amplification will drop. A signal as weak as ten to the minus 15 Watts can be heard. This is, essentially, an S0 signal at 0.1 µV amplitude with only 0.01 µA of current. A regenerative radio can hear AM (carrier with two sidebands) signals, SSB (single sideband) signals, and CW.

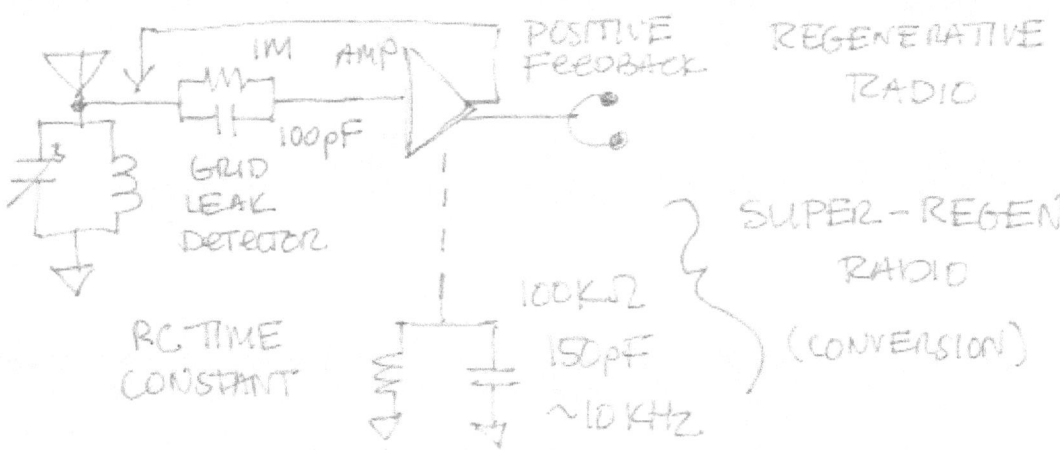

8. SUPER-REGENERATIVE RADIO

A super-regenerative receiver is an oscillator, like a regenerative circuit, that is turned on and off at a supersonic rate. If it is turned on and off at a sonic rate, this tone can be heard. This often gives the radio a characteristic whistle or high-pitched tone. A super-regenerative radio can achieve an amazing 120 dB of gain. This is as much gain as a full-fledged communications receivers, with hundreds of components. However, selectivity is lacking and related to the rate the oscillator is turned on and off at. It also tends to pull to the strongest signals. So, for distance reception (DX) work, the regenerative receiver is still the top single active device radio. In simple terms a super-regenerative radio works by turning an oscillator on and off. The time it takes the self-oscillation to restart depends on the amount of signal present on the tank. To create a super-regenerative radio, take a oscillator circuit and add a resistive-capacitative (RC) time constant (just outside the audio range: ~10+ kHz) in one leg of the active device. A regenerative radio can be turned into a super-regenerative. Capacitor at (*) below inputs RF to the emitter (common base).

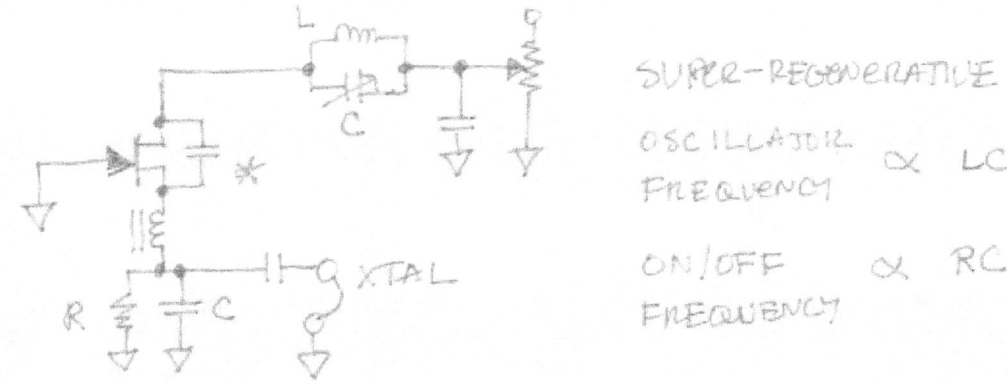

1.4 Oscillators
THE RADIO BUILDING HOBBY
©2015

1. OSCILLATORS: COMMON EMITTER / SOURCE / CATHODE

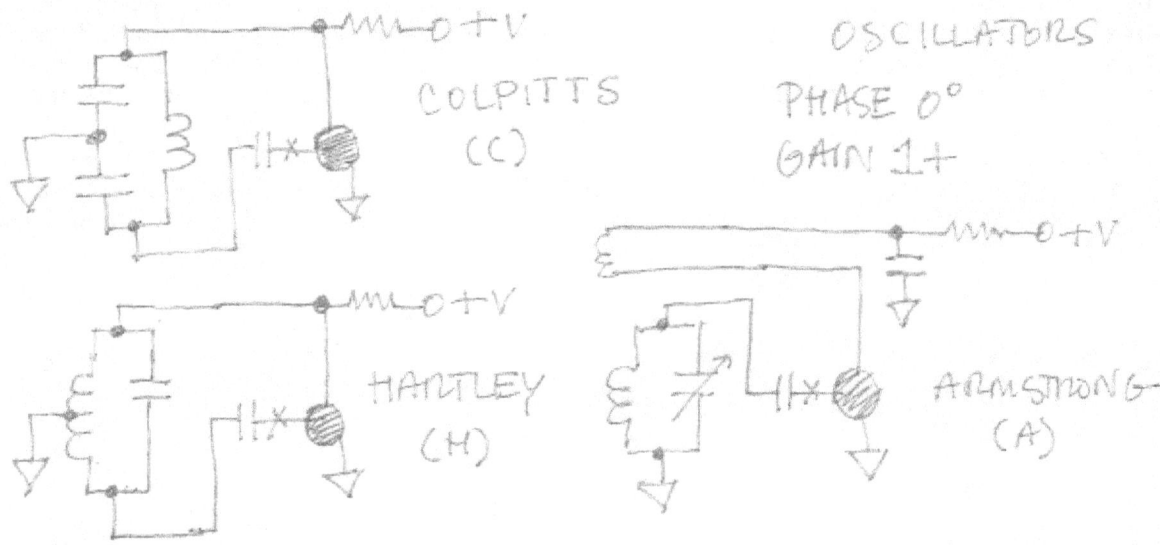

An oscillator is at the heart of every regenerative, super-regenerative, direct conversion (local oscillator), and superheterodyne (local oscillator) radio receiver. Above shows Armstrong, Hartley, and Colpitts oscillator setups in common emitter, common source, or common cathode configurations. The point labeled "X" is for biasing components. This could be as simple as a resistor to positive power in a BJT, negative voltage in a JFET, or a grid-leak detector in a tube.

2. OSCILLATORS: COMMON COLLECTOR / DRAIN / PLATE

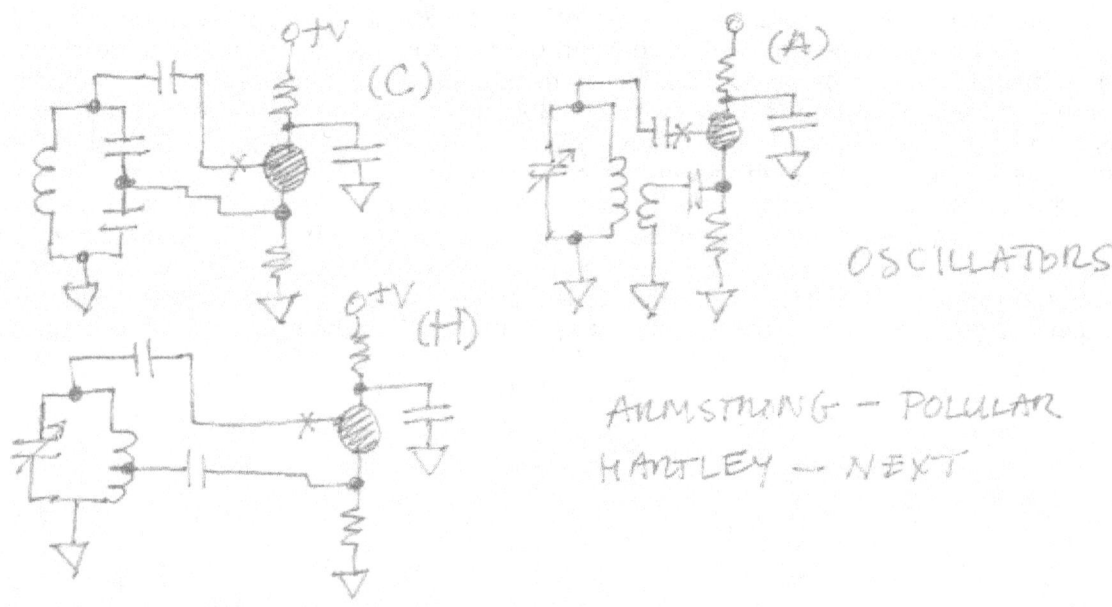

Above shows Armstrong, Hartley, and Colpitts oscillator setups in common collector, common drain, and common plate configurations. Armstrong oscillators are popular in regens. These modes have a voltage gain of one. The Armstrong oscillator's ticker creates voltage gain.

3. OSCILLATORS: COMMON BASE / GATE / GRID

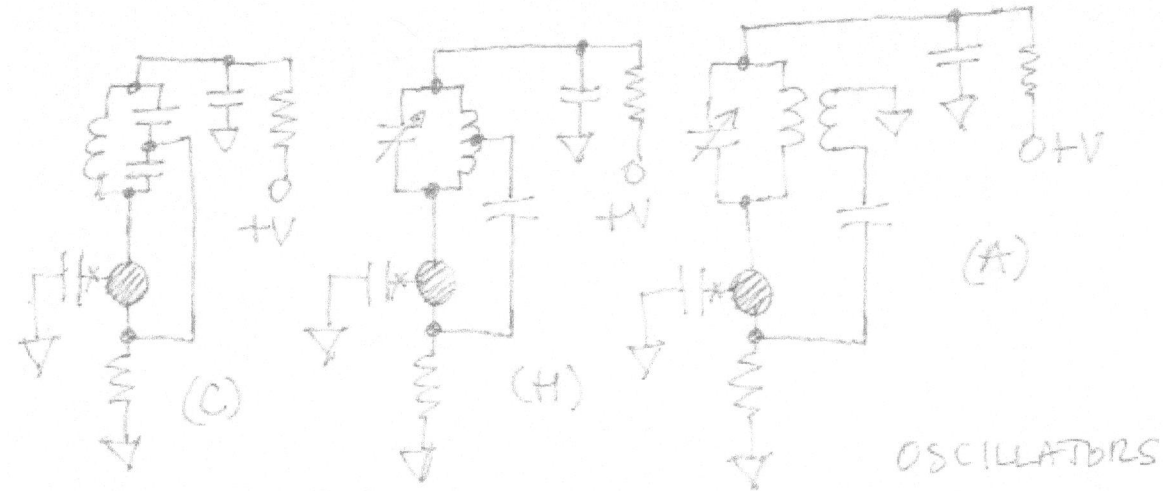

Above shows Armstrong, Hartley, and Colpitts oscillator setups in common base, common gate, and common grid configurations. These modes of operation have a current gain of one.

4. OSCILLATORS

Simple radios deal with three main oscillator types: Armstrong, Hartley, and Colpitts. The difference between them is the way in which regeneration or positive feedback is applied. An Armstrong oscillator using a tickler or wire coil for feedback. Armstrong oscillators can be used with common emitter, common source, or common cathode because the coil can be flipped so that it is now in phase with the input. Voltage gain is inverted in these configurations. If a regen won't regenerate (no sound is heard) the first fix is to swap the tickler coil wires. A Hartley oscillator uses an inductive divider (tank tap) to reintroduce energy. A Colpitts oscillator uses a capacitative divider to reintroduce energy. In general, the Colpitts oscillators are used at higher frequencies than MW and SW. Realize that regenerative radios are just oscillator circuits that are set to a point just before they go into oscillation. They must be able to smoothly go into and out of oscillation. The feedback of energy (the output of the amplifier devices above) must be controllable. This is often done by using a resistor to dump energy before it has a chance to enter the tank. A series capacitor can be used in line with the tickler to control regeneration (Armstrong oscillator). When the capacitor's value is high (~360 pF) its resistance is low and energy is dumped into the tickler. When the "throttle" capacitor's value is low (~30 pF) it presents a load and energy is taken from the tickler and dumped into the capacitor (the highest series resistance gets the most energy). Controlling the supply voltage is a good way to control regeneration and is popular on triodes and JFETs: both are 3-connection, voltage-controlled devices. The screen voltage of a pentode is a good regen control. If a regenerative radio is set to oscillate, the frequency tuned to will directly mix with the frequency of interest and AM signals will be converted down to audio. The regen is now a direct-conversion radio. Selectivity will be enhanced but gain will be reduced. A regenerative radio can also easily be converted into a super-regenerative radio.

AM RADIO RF (kHz)	LO = 1485 ADD	FILTER 455 kHz	LO = 1485 SUBTRACT	FILTER 455 kHz
1010	2495	blocked	475	blocked
1020	2505	blocked	465	blocked
1030	2515	blocked	455	pass
1040	2525	blocked	445	blocked
1050	2535	blocked	435	blocked

Above five AM frequencies are mixed with a LO of 1485 kHz. All the sum frequencies are blocked. However one difference frequency, 1030 kHz passes the 455 kHz filter inside the radio.

5. MIXERS

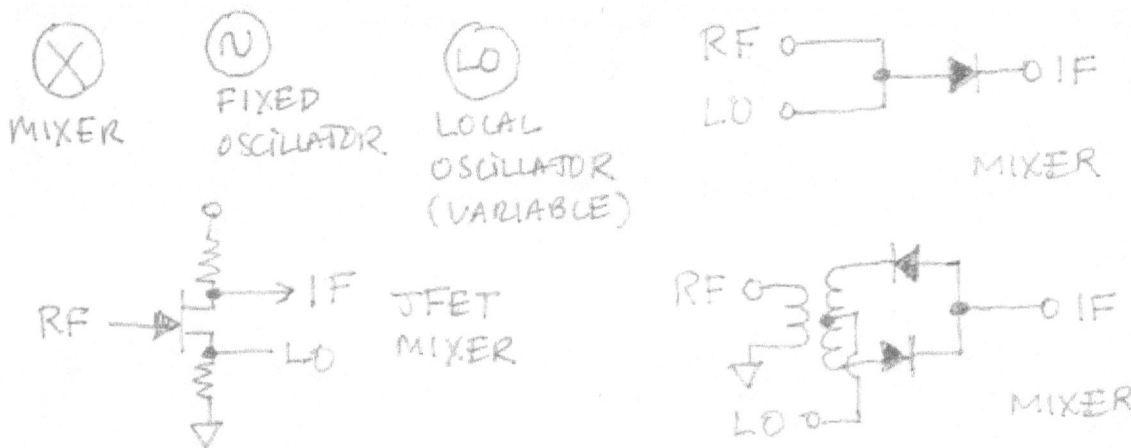

Above shows the symbols for a mixer (X), fixed oscillator (S), and a local oscillator (LO). The local oscillator is variable and is what is changed when turning a radio's tuning dial. Radio frequencies (RF) are mixed with a local oscillator, creating IF or intermediate frequencies. Typically just one frequency is of interest and is filtered by an IF filter. A common IF filter is 455 kHz.

Mixing is not magic, it is physics. When two frequencies are introduced into a nonlinear device such as a diode or diodes (single or balanced mixer, shown above) or amplifier (JFET above), the result is the sum and difference of their frequencies. And also some unmixed original frequencies. It is up to the circuit to filter out the frequency of interest: typically 455 kHz.

There are many types of intermediate frequency (IF) filters including: ceramic (common), mechanical (expensive), crystal, active, RC (resistive-capacitative), inductive "cans", etc.

AM RADIO RF (kHz)	LO = 585 ADD	FILTER 455 kHz	LO = 585 SUBTRACT	FILTER 455 kHz
1010	1595	blocked	425	blocked
1020	1605	blocked	435	blocked
1030	1615	blocked	445	blocked
1040	1625	blocked	455	pass
1050	1635	blocked	465	blocked

Mixing can use either high side or low side injection. In high side injection the LO frequencies are greater than the RF frequencies (see previous page example). The above chart shows low side injection. A LO of 585 kHz is injected and, after mixing, creates many different frequencies. The only one passed by the 455 kHz filter is the frequency that was at 1040 kHz.

AM RADIO RF (kHz)	LO = 595 ADD	FILTER 455 kHz	LO = 595 SUBTRACT	FILTER 455 kHz
1010	1605	blocked	415	blocked
1020	1615	blocked	425	blocked
1030	1625	blocked	435	blocked
1040	1635	blocked	445	blocked
1050	1645	blocked	455	pass

Above shows the LO tuned 10 kHz higher and 1050 kHz being selected by the 455 kHz passing filter. The filter has a bandwidth of about 6 kHz to allow the carrier and both sidebands.

1.5 Real World Amplifiers
THE RADIO BUILDING HOBBY
©2015

1. AMPLIFIERS

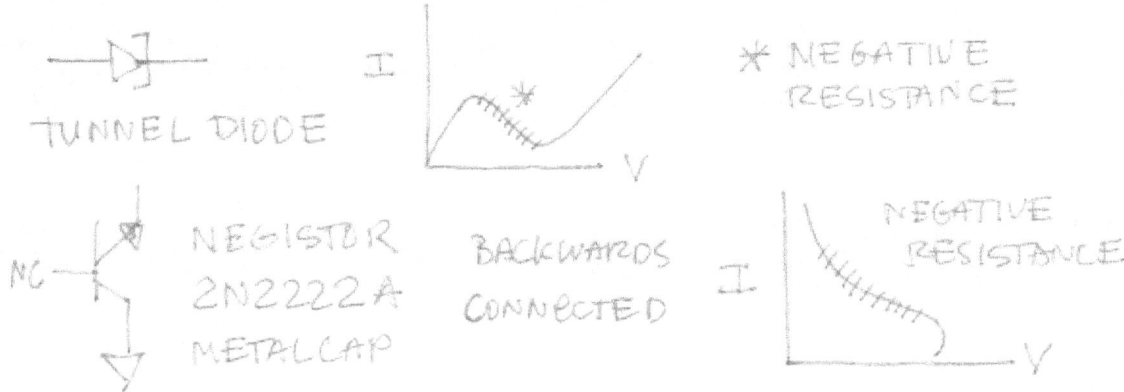

On single active device radios, **sensitivity is achieved using a single amplifier.** That amplifier can be a junction field-effect transistor (JFET), bipolar junction transistor (BJT), triode vacuum tube, or pentode vacuum tube. A JFET has a gate, source, and drain; and BJT has a base, emitter, and collector; and a triode has a gate, cathode, and plate (anode). An amplifier can also be negative resistance, provided by a negistor (metal can 2N2222A) or a costly tunnel diode. Amplification devices only amplify DC. AC signals are amplified by being added to a DC injected (bias) voltage. Coupling capacitors are used to block DC and return the signal to AC. The emitter emits electrons that go to the collector. The source of electrons has them run to the drain.

2. HOW TRANSISTORS REALLY WORK

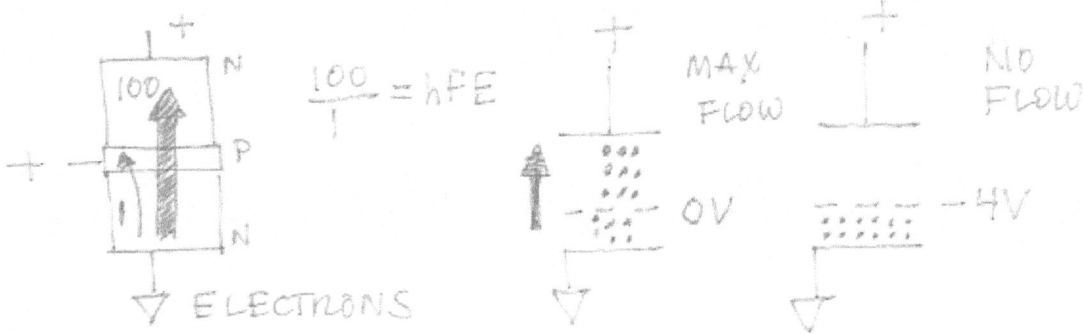

Silicon (better thermal stability) and germanium have a conductivity between that of a conductor (ex. silver or copper) and insulators (ex. rubber or glass). Alone, silicon acts more like an insulator. Silicon (and germanium) contains 4 outer shell (valence) electrons that can be shared with other silicon atoms. Boron has three outer shell electrons (electron deficient, extra hole) and is added to make the "P" (positive charge carrier) type silicon. Phosphorous has five outer shell electrons (electron rich, extra electron) and is added to make the "N" type (negative charge carrier) silicon. An NPN transistor has a thin and lightly p-dope region sandwiched between two thick n-doped regions. One n-doped region leads to positive power; the other n-doped region leads to ground. The p-doped region goes to the positive supply, through a large resistance. At least 0.7 Volts must be present because the base-emitter form a diode. Electrons, from ground (negative supply) travel across the n-doped region. They are attracted to the p-doped region. However, being thin, for every 1 electron that actually makes it to the p-doped region's terminal, 100 electrons end up going past the thin p-doped region and end up in the n-doped region attached to the positive power supply (ex. through a load). This ratio of 1:100 represents a current gain of 100.

3. HOW VACUUM TUBES REALLY WORK

See the above picture, right side. A triode is made up of a filament (heater, hooked to A+ power) that gets hot and boils off (emits) electrons. This is called the cathode. Some tubes have a physical cathode plate. Electrons, with a negative charge, are attracted to the plate (anode), with its positive charge (hooked to B+ power). Electrons fly through the vacuum (an absence of air, lacking N_2 and O_2) from cathode to anode. This alone is a diode or rectifier, with electrons only flowing in one direction. There is maximum plate current at zero grid voltage. A grid is located between the anode and cathode. The more negative the grid, the less the cathode electrons are attracted to the anode: hence, less current. The electrons of the grid repel the electrons of the cathode. At some negative gate voltage the electrons are no longer attracted to the plate and the flow of electrons is cut-off. The closer the grid is, physically, to the cathode, the higher the transconductance (Gm) of the tube. The 6GK5 is a high Gm tube due to frame-grid construction.

4. HOW JFETS REALLY WORK

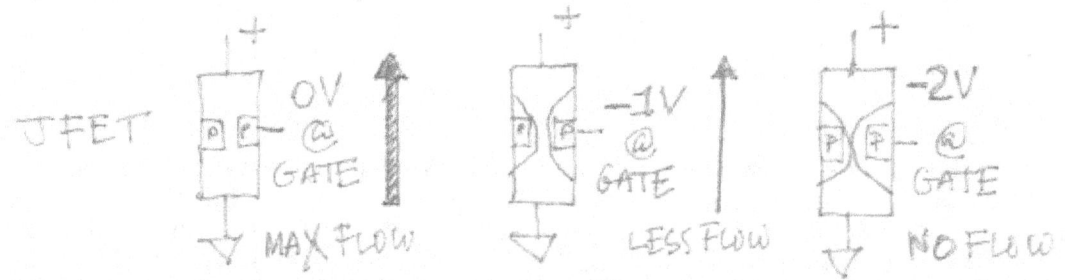

An N-channel JFET works similar to a vacuum tube. Electrons flow from ground to the source lead through the n-channel to the drain lead to the positive supply. Maximum current occurs at zero gate voltage. As gate voltage is decreased to minus one volts, the depletion zone increases and the current is reduced. At a gate voltage of minus two volts there is no flow. The amount of negative voltage required to shut off current flow varies from -2V to -6.5V in a J310 JFET.

5. BIPOLAR JUNCTION TRANSISTOR: MPSA18

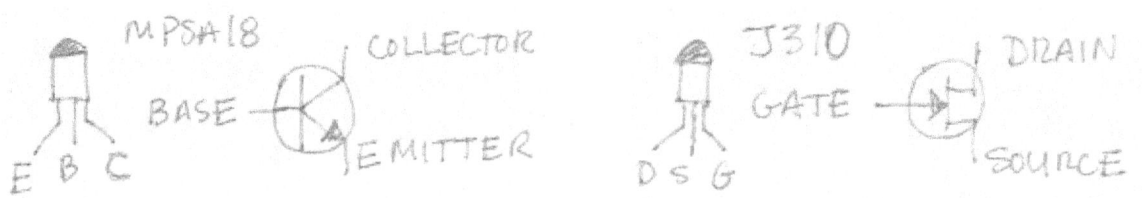

The **MPSA18** is an good transistor (BJT) or for single active device radios. It is classified as a "General Purpose Amplifier". A BJT is a current controlled device. Its important specs will be noted. The absolute maximum ratings are: 45 Volts (V_{CEO}) and 100 mA (I_c, collector current). Only 1.5 Volts and 1 mA of current are needed for a simple radio. Minimum DC current gain (hFE, also called beta) is 400 at 10 μA I_c and **500** at 10 mA I_c. The MPSA18 has high current gain that increases with collector current. The maximum collector-base capacitance (C_{CB}) is 3.0 pF. The maximum emitter-base capacitance (C_{EB}) is 6.5 pF. Capacitance can pass RF, un-amplified. The minimum current gain bandwidth product (f_T) is 100 MHz. At 100 MHz current gain is ~1. Maximum noise figure (NF) is good at 1.5 dB. This transistor works well for MW and SW radio projects.

Also noteworthy is the **9018** or **J8** (Chinese), an "AM/FM amplifier or local oscillator for a FM/VHF tuner". It has a current gain bandwidth product (f_T) of 1.1 GHz (typical and 700 MHz minimum). And a hFE of ~100. It can be operated up to 30 Volts (V_{CEO}) and at a collector current of up to 50 mA (I_c). This transistor is better for FM applications. Output capacitance is 1.3 pF (typical). The **MPSA14** is a 30V darlington transistor with an hFE of 10,000 (min) and a Vce of 1.5 Volts.

6. JUNCTION FIELD EFFECT TRANSISTOR: J310

See the picture above, right side. The J310 is a good transistor JFET for single active device radios. It is classified as a "N−channel (Depletion) FET VHF/UHF Amplifier". A JFET is a voltage controlled device. The important specs will be noted. The maximum ratings are: 25 Volts (V_{DS}) and 10 mA of gate forward current (I_{GF}). The gate is normally reverse biased: this accounts for the high input impedance. If forward current exceeds 10 mA the device can be destroyed. Static can also destroy the gate. The minimal gate-source cutoff voltage ($V_{GS(off)}$) is -2.0 Volts. This is the most critical specification. Its maximum is -6.5 Volts. One problem with JFETs is that their individual characteristics vary greatly. The zero gate voltage drain current (I_{DSS}, or maximum current that occurs when the gate voltage is zero) is from 24 mA to 60 mA (notice the variation). The common-source forward transconductance (g_{fs}) is 8,000 to and 18,000 µmhos (8 to 18 mS). Common gate power gain at 10V, 10mA, and 100 MHz is 16 dB. Gate-drain capacitance (C_{gd}) is 1.8 pF. Gate-source capacitance (C_{gs}) is 4.3 pF. This is a very popular JFET transistor for radio construction. The J310 is a symmetric device: the drain and source could be swapped but it is not advisable.

The **J309** is another good RF JFET; but is hard to find. It has an Idss that ranges from 12 to 30 mA. Its Vgsoff varies from -1 to -4 Volts. And its transconductance can be from 10 to 20 mS.

The **MPF102** is a commonly used JFET with poor performance. It has an Idss (drain current at zero volts of gate) of 2 to 20 mA. Its Vgsoff varies from an unspecified minimum to -8 Volts maximum (meaning a good 9V battery will just shut it off). Its transconductance is 2.0 to 7.5 mS.

The J310, J309, and MPF102, when lying flat-side down, pins towards you, have their gate on the left, source in the middle, and drain on the right. Try not to touch or handle the gate.

7. TRIODE VACUUM TUBE: 6GK5

PIN 1 CATHODE
PIN 2 GRID
PIN 3 HEATER
PIN 4 HEATER
PIN 5 PLATE
PIN 6 INTERNAL SHIELD
PIN 7 CATHODE

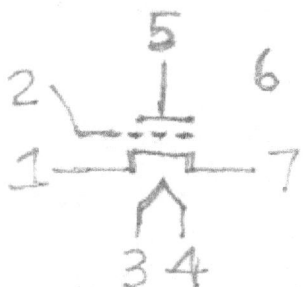

The 6GK5 is a frame-grid, triode, vacuum tube for single active device radios. It is classified as a "VHF high-GM triode RF Amplifier". A triode is a voltage controlled device. The important specs will be noted. The G6K5 plugs into a 7-pin miniature base and can be mounted in any position. The heater uses 6.3 Volts at 180 mA (6GK5). Or 2.8 Volts at 450 mA (3GK5). Capacitances are: 0.52 pF grid-to-plate, 2.5 pF heater-to-cathode, 5.0 pF input, and 3.5 pF output. A pF is the same as a µµf or "mickey-mike". Maximum plate voltage is 200 Volts: radio does not require this much voltage. Maximum cathode current is 22 mA. Negative grid voltage is 50 Volts. Grid resistance is 1.0M ohm. Class A1 Amplifier, typical characteristics are: 135 Volts of plate, -1 Volt of grid voltage, 78 amplification factor, 5400 ohms of plate resistance, 11.5 mA of plate current, a whopping 15,000 µmhos transconductance (gm) or 15 mS. This is reduced to 1500 µmhos at -2.5 Volts (gate) and 150 µmhos at -4.2 Volts (gate). The noise figure is a respectable 4.7 dB.

Using techniques talked about elsewhere in this book, important parameters were determined from graphs of the 6GK5's "average plate characteristics". At 30 Volts of plate, plate resistance is 5,330 ohms, transconductance is 9,000 uS, and plate current is 6.5 mA. At 20 Volts of plate, plate resistance is 4,660 ohms, transconductance is 7,300 uS, and plate current is 5.2 mA. At 10 Volts of plate, plate resistance is 2,770 ohms, transconductance is 6,200 uS, and plate current is 3.5 mA. Transconductance is high, even at low (safe) plate voltages. This is an ideal tube for high-performance single active device radios. The 6GK5, 3GK5, and 2GK5 are similar tubes with different heater voltages. The transconductance of this tube, at 30V, is comparable to a JFET.

8. PENTODE VACUUM TUBE: 6418

The 6418 is a filament-type (no cathode lead), subminiature, pentode vacuum tube. It is classified as a power amplifier for portable equipment. A pentode is a voltage-controlled device. Important specs will be noted. The filament (heater) needs 1.5 Volts at **10 mA** (AC or DC, any position). The dot is lead number one. No socket is needed. Maximum plate and grid #2 voltage is 30 Volts. Maximum cathode current is 0.5 mA or only 500 µA. Typical values are: plate and grid #2 voltages of 22.5 Volts, grid #1 voltage of -1.2 Volts, zero-signal plate current is 240 uA (maximum), zero signal grid #2 current is 60 uA, peak audio grid #1 voltage is 1.2 Volts, transconductance is 300 umhos, plate resistance is 0.42M ohms, load resistance is 100k ohms, distortion is 12%, and power usage is 2.2 mW. Pentodes have two extra grids called a suppressor grid (closest to the plate) and screen grid (next closest to plate). Grid #2 on the 6418 and 6612 is the screen grid: it reduces grid to anode capacitance. The suppressor grid is internally connected to the filament (ground, pin 3). The 6418 and 6612 are used in common cathode configuration, only. The screen grid reduces the Miller effect (capacitance) and helps increase plate resistance.

PENTODE 6418 6612

PLATE 1.

SUPPRESSOR.

G2 SCREEN GRID 2.

G1 CONTROL GRID 4.

FILAMENT

3, 5.

DOT = LEAD 1 (PLATE)

"3" = GROUND

9. PENTODE VACUUM TUBE: 6612

See picture above. The 6612 is a filament-type (no cathode lead), **shielded**, subminiature, pentode vacuum tube. It has internal shielding to prevent radio leakage. It is classified as used in RF amplifiers. A pentode is a voltage-controlled device. The filament (heater) needs 1.5 Volts at **80 mA**. The red dot is lead one. No socket is required. Maximum plate and grid #2 voltage are 50 Volts. Maximum cathode current is 5.5 mA. Typical values are: plate and grid #2 voltage of 45 Volts, zero-signal plate current is 3 mA (maximum), peak audio grid #1 voltages are 1.2 Volts, transconductance is 3000 umhos, and -2.6 Volt cut off voltage. This is a hotter tube than the 6418 with ten times the transconductance but it uses eight times the heater current. This tube is meant to be used only as a common cathode amplifier. Note: A vacuum tube's anode is also called the plate. The 6612 has a plate resistance (Rp) of 180k ohms and a transconductance of 3 mS.

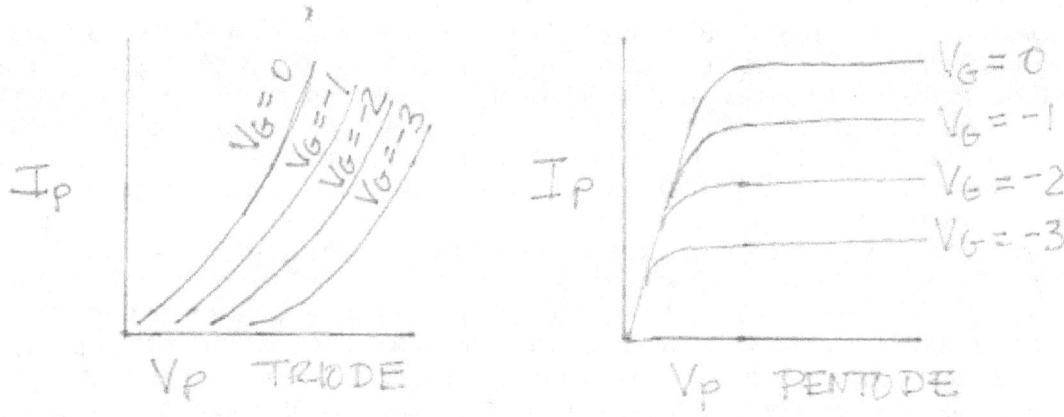

The plate current versus plate voltage curves of a triode and pentode are shown above. The pentode curve looks similar to an n-channel JFET device. In a regenerative radio the triode plate voltage may be used to control regeneration. In a pentode the screen grid is often used.

1.6 Amplifier Modes of Operation
THE RADIO BUILDING HOBBY
©2015

1. ACTIVE DEVICE MODES

An active device can provide gain in three distinct modes. In a single active device the radio (RF) and audio (AF) signals can use different paths in-to and out-of the active device. These modes are often chosen because of their input and output impedances. See the summary below. The circle below represents the device (BJT, JFET, triode). For example: the top arrow shows that input to the base of a BJT has medium input impedance and medium output impedance through the collector. The emitter is not used (not an input or output): this is common emitter. This has both voltage (V) and current (I) gain for high power gain. Note that the input impedance of both a JFET and triode are high. What the diagram below represents will be explained in detail below.

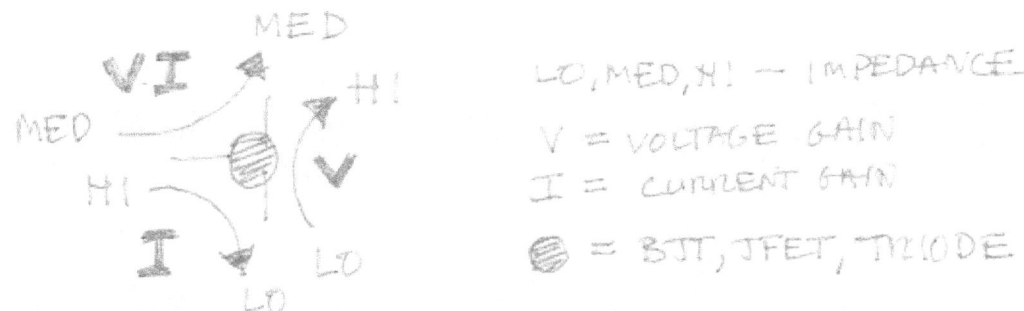

2. COMMON EMITTER, COMMON SOURCE, COMMON CATHODE

Common emitter (BJT), common source (JFET), and common cathode (triode, pentode) amplify both current and voltage. Output phase shift is 180 degrees (inverted). Regenerative feedback must inverted, often by using an inductor. Direct feedback will cause degeneration or weakening of the signal. Input impedance is moderate in the BJT but high in a JFET or tube. Output impedance is moderate. Power gain is the highest in this mode. This is the most common mode used for any active device. It is subject to the Miller effect, wherein the voltage gain causes an increase in the equivalent input capacitance (C = C x gain). A resistor in combination with a bypass capacitor can be used on the emitter, source, or cathode to allow common emitter, common source, or common cathode configuration at audio or radio frequencies (or both). The resistor is used to increase gain stability. This mode often uses a resistor and bypass capacitor (small for radio, larger for audio). In a BJT, the purpose of the resistor is to get voltage gain away from being dependent on internal emitter resistance (re), which is unstable. The BJT's **re** varies with temperature and collector current. The bypass capacitor increases the voltage gain at radio or audio. It does this by reducing the resistance at these frequencies from the emitter to ground.

3. COMMON BASE, COMMON GATE, COMMON GRID

Common base (BJT), common gate (JFET), and common grid (triode, pentode) amplify only voltage. Current gain is ~1 (one). Phase shift is 0 degrees: regenerative feedback can be directly applied to the tank. Input impedance is low (good for accepting incoming RF); output impedance is high (good for driving a voltage-controlled amplifier or feeding a tank). Power gain is the lowest in this mode. Voltage gain is increased by resistance above the device (ex. collector load). And reduced by resistance below the device (ex. emitter resistor). The resistor in the emitter, source, or cathode leg should be about two to ten times the incoming impedance so that the signal is not lost to ground; and enters the emitter, source, or cathode. This mode of operation is commonly used for RF voltage amplifiers due to the low impedance of the incoming RF wave (377 ohms).

4. COMMON COLLECTOR, COMMON DRAIN, COMMON PLATE

Common collector (BJT), common drain (JFET), and common plate (triode, pentode) amplify only current. Voltage gain is ~1 (one). Phase shift is 0 degrees: regenerative feedback can be directly applied to the tank. Input impedance is high (will not load a tank); output impedance is low (great for driving real-world devices like a tickler or phones). Input impedance is very high in the JFET and vacuum tube. Power gain is low in this mode for the BJT but moderate for voltage-controlled devices (JFET and tubes). This mode goes by a second name. In the BJT it is called an emitter follower. In the JFET it is called a source follower. And in the vacuum tube it is called a cathode follower. Current gain on a BJT is its h_{FE} value: these often run from 50 to 500. Current gain on a JFET or vacuum tube is very high because the large input impedance means very few electrons are actually flowing. Current gain here is "infinite" but, in practice, the power gain is not.

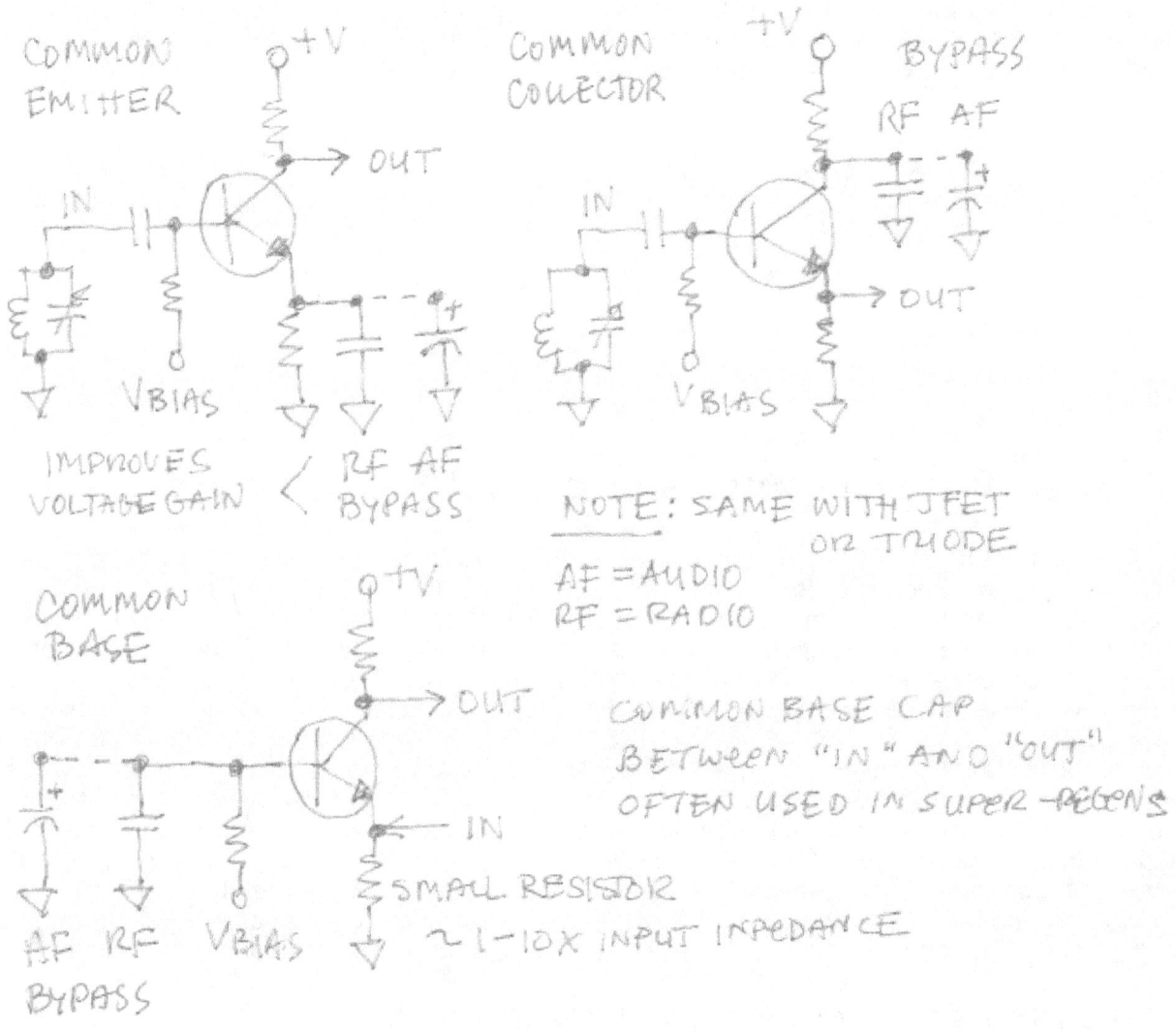

5. GRAPHICS

Study the graphics above. Notice the way that bypass capacitors allow DC resistances to remain in place but not effect the way that radio (RF) or audio (AF) are being amplified. In single active device radios it is critically important to understand all three modes of each device type. Note that radio or audio only need to be presented back to ground after going through the active device. There are no positive and negative poles in alternating current (radio or audio waves).

6. TRANSISTORS AS SWITCHES

A transistor can be made into an ON and OFF switch. In the modes between on and off, a transistor is used as an amplifier. The amplifier will fall into class A, class B, or class AB operation, depending on its biasing. A BJT is off when ground (or up to 0.5V, Si) is applied to its base. A BJT is on when a positive current is applied to its base. This current is <1% of load current. A JFET is off when a negative voltage is applied to its gate. The exact number of negative volts to turn the JFET transistor off will vary greatly. A JFET is on when ground or zero volts is applied to its gate.

7. TRANSCONDUCTANCE

Transconductance (g_m) is the change in current in the output divided by the change in voltage in the input. Transconductance is measured in Siemens or mhos. The unit mho is the word "ohm" spelled backwards. And the symbol for transconductance is an upside-down ohm symbol. A JFET or vacuum tube is a voltage controlled device. A voltage controls the drain or plate current. The MPF102 is a low-performance JFET with a transconductance of 2,000 to 7,500 μmhos or μS. The J310 is a high-performance JFET with a spec of 8,000 to 18,000 μS. The 6GK5, a frame grid triode, has a spec of 15,000 μS. This is a 1960's tube that can compete with a modern JFET. The 6612 has a spec of 3,000 μS and the 6418 a spec of 300 μS. Although BJT transistors are current-controlled devices (a small base current controls a large collector current), they have very high transconductance. The transconductance of a BJT, at room temperature, is equal the the collector current divided by 26 mV. This means that at only 1 mA, the transconductance is 38,400 μS or 38.4 mS. At 10 mA of collector current, transconductance is a whopping 384,000 μS or 384 mS. The problem with a BJT is it's low input impedance: which is dependent on its current gain.

8. DURABILITY

Vacuum tubes and BJT transistors are robust devices. A JFET has a relatively delicate gate. A MOSFET is a notoriously weak device that can be destroyed easily with improper handling. It has a very delicate gate. Static can easily destroy a MOSFET (ex. VN10KM, N-MOSFET, 300 mS). Vacuum tubes are very forgiving and durable devices and very safe when using one, two, or three 9 Volt batteries, in series, for the plate voltage. Vacuum tubes also go into shutoff much smoother than silicon devices. They tend to distort instead of abruptly turning off. It is very important to have a load present or excess current from the battery could destroy a semiconductor. This means that from ground to the emitter/source and/or from plus power to the collector/drain there must be some DC resistance. A Bogen T-725 transformer (1420 ohms at DC) suffices as it will only allow 6.2 mA of DC current to flow from a 9 Volt battery. Or 12.4 mA from two 9V batteries. With a 1.5 Volt battery only 1.1 mA of current will flow. Without the Bogen, about 1000 ohms in the circuit will limit current to 9 mA at 9V Volts or and 1.5 mA at 1.5 Volts. At 9 Volts a J310 JFET will need at least 500 ohms in the source or drain leg to limit current to 18 mA. At 1.5 Volts an MPSA18 BJT will need at least 30 ohms in the emitter or collector leg to limit current to 50 mA. For reason described below, the JFET ideally needs **at least** a 9V battery supply; while a BJT can easily run on 1.5 Volts.

1.7 Amplifier Experimentation
THE RADIO BUILDING HOBBY
©2015

1. THEORY

There is a considerable amount of math and theory behind amplifiers: BJT, JFET, triode, and pentode. A person could spend a lifetime learning it all. One of the best ways to learn is to build circuits like the ones in this manual and then alter component values. It is actually easier to solder in a component and listen to hear the changes than to go through all the calculations. And what is calculated does not always match reality. Electrons do not care about models or formulas. Electrons are always ready to teach us lessons. It took a 12 page discussion on a radio forum, with radio experts, to determine how the Hellenedyne receiver was detecting. And in the process it was determined that a brilliant radio designer, Sir Douglas Hall, may be in error with how one of his most prominent circuits worked as a detector: the Spontaflex. And his circuit appeared in 1964.

2. LOW COST EXPERIMENTATION

An MPSA18 costs 7 cents in bulk. A JFET costs 20 cents in bulk. A 6418 costs $1 in bulk. A 6612 costs $4 each. And a 6GK5 costs $7 each. In 14 years of soldering, I have managed to destroy one tube. This occurred by applying 9 Volts to a 1.5 Volt heater (it sure did glow bright). Also destroyed were about 10 JFETs via puncturing a hole in their gate. Tubes and BJT transistors are pretty hard to destroy at the voltages used for radio. A JFET can be destroyed more easily but they only cost 20 cents. Use low voltage and batteries for safety and have fun experimenting.

3. MODES OF OPERATION

If more load (at radio or audio, independently) exists above the collector, drain, or plate, then the circuit leans to being common emitter, common source, or common cathode. A resistor with a bypass capacitor (in parallel) can be used in this mode since a small capacitor will allow RF a path to ground. And a large capacitor will allow both audio and radio a path to ground. If more load exists below the emitter, source, or cathode, then the circuit leans to being common collector, common drain, or common plate. Again, a small (RF) or large (RF and audio) capacitor can bypass these to ground (above the device) and setup this mode. If a large load exists above and a small load exists below the device, then the circuit leans to being common base, common gate, or common grid. Input will be into the emitter, source, or cathode. The real key is to determine: the input, the output, and what is grounded. Using an antenna and tank, experiment with the different oscillators and modes of operation of each device: BJT, JFET, triode, and pentode.

4. LIMITING CURRENT

While experimenting, it is critical that some direct current load exists either above (collector, drain, plate) or below (emitter, source, cathode) the device. Each device can only handle so much current. A load means a resistor or wire (ex. audio transformer primary) that is 500 or more ohms at 9 Volts. This limits current to 18 mA. Even better, using 1000 ohms limits current to 9 mA. Check the amplifying device's specifications. An MPSA18 can handle 100 mA of collector current. A J310 can flow 24 mA to 60 mA at maximum (when the gate voltage is zero). Vacuum tubes are typically not going to be damaged at low plate voltages. If a Bogen transformer is used, with its 1420 ohms of DC resistance, it will be next to impossible to destroy any of these devices via too much current. And the 1.5V to 9V voltages used are well under what each device is rated at: MPSA18 (45 Volts), J310 (25 Volts), 6GK5 (200 Volts), 6612 (50 Volts), and 6418 (30 Volts). In terms of current: an old-time radio vacuum tube needs as much as 25 Volts of plate to produce 1 mA of plate current. A 6418 needs 22.5 Volts to create 240 µA of plate current. A 6612 needs 30 Volts to create 3 mA of plate current. And a 6GK5 needs 10 Volts to create 3.5 mA of current. The current needed for a radio receiver using sensitive phones is only about 1 mA.

5. JFET GATES

A J310 JFET can be damaged if its maximum forward gate current (I_{GF}) of 10 mA is exceeded. If positive voltage is applied to the gate of an N-channel JFET, the junction is forward biased. Current can now flow from gate to drain (or source) and, if high enough, destroy the device. These devices can be damaged by static. Do not walk across a carpet in winter and then touch a JFET. A positive aspect of a JFET, shared with vacuum tubes, is their high input impedance. This keeps them from loading a tank circuit. A JFET with zero volts on its gate allows maximum flow of electrons from source to drain. As the negative voltage increases on the gate, this flow is shut down. This is because the depletion area around the p-doped region is increased in size.

6. DIGITAL MULTIMETER USAGE

A multimeter can be used to assess voltage, resistance, and current. It is safer to calculate current because a meter can easily be damaged in that mode. The gate or grid bias voltage can be measured on a JFET or tube, respectively. The voltage at the resistor below the source or cathode can be measure on a JFET or tube. This will measure the bias voltage. The drain or plate current can be calculated by measuring the load resistance and the load's voltage drop. The emitter voltage (~10% of supply for stability) can be measured when a resistor and bypass are used. The collector voltage can be measured (~45% of supply for full swing). Optimal resistances can be determined via testing and listening. Then a standard resistor can be dropped into the circuit and the potentiometer removed. This is not the way they teach in college; but it make experimenting more enjoyable. On the web, there are good calculators for transistor and tube amplifiers.

7. CRYSTAL EARPHONE USAGE

Use earphones to assess how well a circuit is working. Nothing beats success. If the audio is clipping, the bias could be wrong. If the audio is missing there may be no regeneration. Between usage of a simple multimeter and actually listening using a crystal earphone, there is really no need to use a $2000 oscilloscope. If there is amplified radio, it can be heard by using a diode attached to a crystal earphone to ground. The crystal earphone blocks direct current, acts as an RF path to ground, and will smooth the detected (turned to audio) radio. Ears can be used an an oscilloscope. A regenerative radio can hear almost any signal a communications receiver can hear.

8. REGENERATIVE CONTROL

The Globe Patrol Junior and Dee/Mitch-dyne II both use a 10k-ohm potentiometer attached to ground and +9 Volts. The wiper is attached to the MPSA18's base via a 470k-ohm resistor. This allows altering the base hole current such that from 0.00 to 0.85 mA of current flows in the collector. The 470k-ohm resistor minimizes any loading, while allowing compete transistor control.

The Heliosdyne controls regeneration by altering drain voltage. A 10k-ohm potentiometer is attached to ground and +9 Volts. The wiper is attached to an audio transformer with 1420 ohms of DC resistance. This setup is reminiscent of old triode vacuum tube regenerative radios, that are controlled via plate voltage. A large source resistor sets the JFET bias to be near its cut-off.

The QRP tube super-regenerator and Oscillodyne control super-regeneration using the screen grid of a pentode. This grid can totally control plate current. The under-tank capacitor and resistor combination cause super-regeneration. The resistor is a low value of only 470k-ohms and drains electrons off the grid at a faster rate than the typical 1M ohm resistor. The other feature that causes super-regeneration is the large tickler coil. This disrupts the tube in a rhythmic fashion.

The Angelodyne, JFET Hellenedyne, and triode Hellenedyne control the JFET or triode using source or cathode biasing resistors. They are variable so that they can be optimized for each individual JFET or triode. The regeneration control is via a tickler. Again, the voltage used for biasing can be measure via a multimeter. The next section will discuss biasing in more detail.

1.8 Amplifier Biasing
THE RADIO BUILDING HOBBY©2015

1. COUPLING CAPACITORS

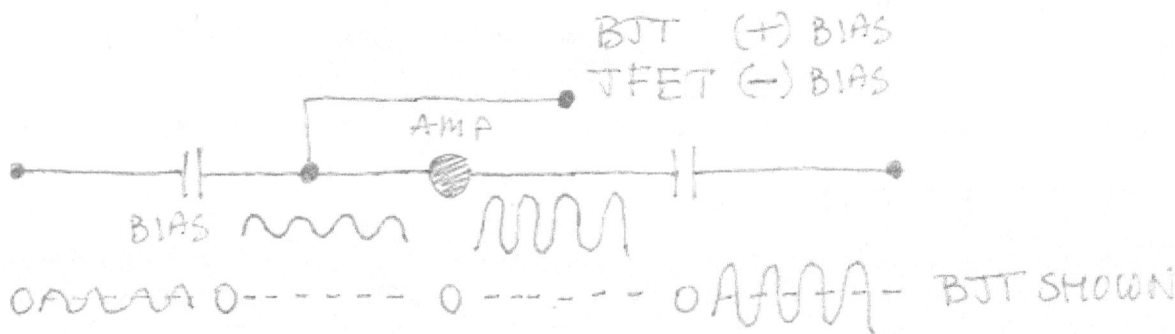

Shown above is the purpose of the two coupling capacitors, one before and one after an amplifier. An tiny alternating current (AC) signal passes the first capacitor with little resistance. This capacitor prevents the bias voltage (positive in a NPN BJT transistor, negative in an n-channel JFET transistor, negative in a triode, and negative in a pentode) from leaking through the signal source to ground. The AC signal now rides on top of a DC bias voltage (above shown as positive or above zero). An amplifier only amplifies direct current (DC) signals. For example: directly feeding an AC signal into an NPN transistor will only cause amplification of that part of the signal that is above 0.7 Volts or 700 mV that is positive. The rest of the signal would be lost. In radio, a strong signal S9+10dB is only 160 uV or 0.160 mV. Without biasing the amplifier would lose all the radio signal. After the amplifier (see above) the signal is now amplified but still riding on DC. The second capacitor removes the DC signal and returns the, now amplified, signal back to being AC.

2. BJT BIASING

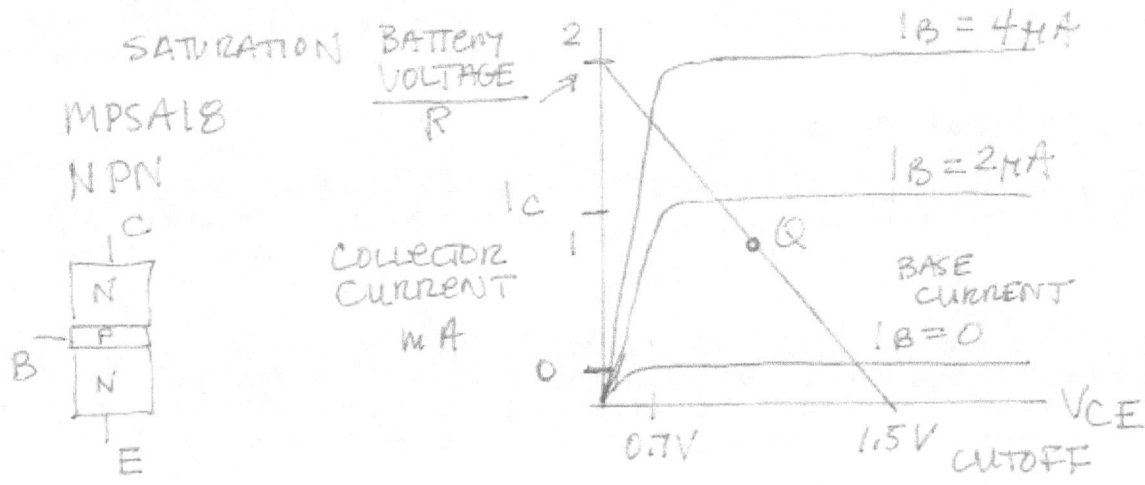

The graph has an x-axis of collector-to-emitter voltage and a y-axis of collector current. The three curved lines plot a base current (Ib) of 0, 2, and 4 µA. The output or collector current is Ib time 500 (hFE, beta, or current gain) or 0, 1, and 2 mA (y-axis). This collector current does not occur until Vce is above about 700 mV. A load line is drawn with a "Q" point. If this Q point is near cutoff (here 1.5 Volts or the supply voltage), then the amplifier is in class B mode (only half of the signal is amplified). At the Q point shown it is in class A mode (the entire signal is amplified). Cutoff is the supply voltage because at cutoff the voltage at the collector equals the supply voltage. This occurs because the transistor is off and can be thought of as an open circuit. The supply voltage flows through whatever collector resistor is present and is present at the collector because it is essentially not hooked to anything else. The other end of the load line is in the upper left corner of the graphic. It is the maximum current or supply voltage divided by the the collector resistor's value (Ohm's law). The Q point is set along this line by setting the base current. If the base current is low it is near cutoff (open switch). If the base current is high it is near saturation (closed switch). An AC signal, such as radio or audio, moves above and below the Q point and is amplified by the transistor. If close to cutoff or saturation, only half of the signal may be amplified.

A BJT must be biased properly to function. The base's P-region must be attached to some positive voltage to work. The base's P-region and emitter's N-region form a diode. There must be 0.7 Volts or 700 mV before any current flows in the base (and hence in the collector and emitter). Although the MPSA18 can handle 100 mA of collector current, a radio only needs a small current, about 1 mA, to function. A MPSA18 has a gain of 500. So 500 times less current is needed at the base to get that 1 mA of current at the collector: 2 µA. The supply voltage of 1.5 Volts minus the 0.7 Volt base-emitter voltage drop (Vbe) results in 0.8 Volts. The current in the base (that sets the Q point) is equal to this 0.8 Volts divided by both the base resistor plus 500 (the hFE) times 26 (the emitter resistance at 1 mA). The value of 500 times 26 or 13k ohms is often insignificant compared to the base resistor and it is omitted. The simple calculation uses 0.8 Volts and 2 µA in Ohm's law.

Take an MPSA18 hooked to 1.5 Volts with a theoretical load of only 1 ohm in the collector and emitter circuits. The MPSA18's maximum collector current of 100 mA is not reached unless a tiny 3.5k ohm base resistor is used. The designs in this book use a 470k ohm resistor attached to a pot that is attached between ground to +1.5 Volts. Collector current is, at most, 0.85 mA or 850 µA. Both the Globe Patrol Junior and Dee/Mitch Dyne II use this setup. A 100k, 200k, and 400k ohm base bias resistor results in a maximum of 4.0 mA, 2.0 mA, and 1.0 mA of collector current (respectively). With a current gain of 500, not much base current is needed to make the MPSA18 transistor work as an amplifier. By using a potentiometer, the Q point can be changed from cutoff (under 0.7V) to near saturation. In class B operation, where the Q point is near saturation or cutoff there is clipping of part of the signal. A sweet spot in the middle of acceptable values may be found. The resistance can then be measured with a a multimeter. And then just added to the circuit. This is a way of determining values without using any math. This bias method is called current bias and is shown below as (1). Using a pot, this method is acceptable. However, using a resistor, this method is unstable because it is highly dependent on re (internal emitter resistance) which varies with temperature and Ic. Method (2) below or collector feedback bias is more stable.

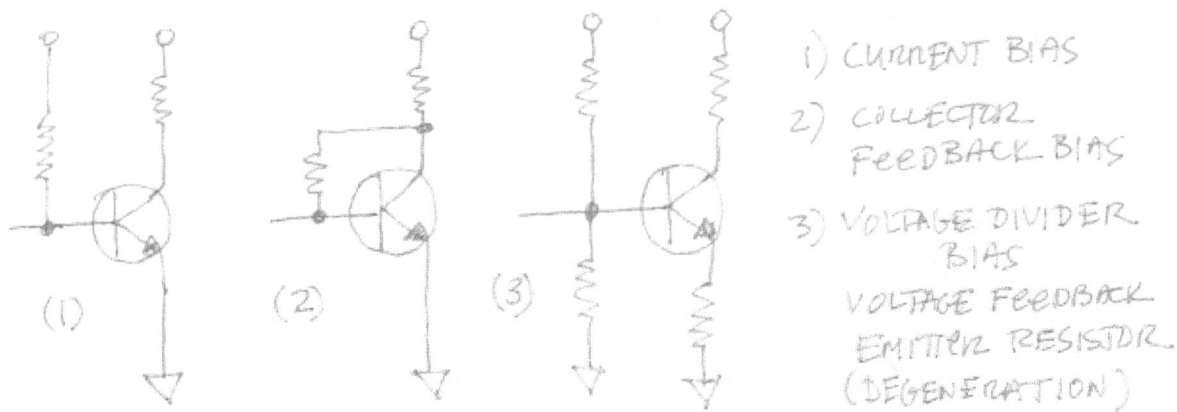

1) CURRENT BIAS

2) COLLECTOR FEEDBACK BIAS

3) VOLTAGE DIVIDER BIAS

VOLTAGE FEEDBACK EMITTER RESISTOR (DEGENERATION)

A third method (3) is called voltage divider bias. This method uses a resistor ladder at the base to pour current into the base. However the base current is controlled by the emitter resistor. A good value to shoot for is 10% of supply voltage on the emitter resistor. At 1.5 Volts of supply, 10% is 0.150 Volts. If a collector current of 1 mA is desired, then by Ohm's law the emitter resistor is 150 ohms. If the fullest swing is wanted at the collector, the collector resistor should see 45% (half of the remaining 90%) of the supply voltage or 0.675 Volts. Using Ohm's law the collector resistor should be 675 ohms. Notice the collector resistor is 4.5 times the emitter resistor's value. The emitter resistor should be bypassed by a capacitor (small for RF, large for audio) to increase voltage gain. The base voltage is 0.150 Volts (emitter resistor) plus 0.700 Volts (base-emitter voltage drop) or 0.850 Volts. If the top ladder resistor is 100k ohms (equation: Vb = R2/(R1+R2) * supply), the bottom will be 131k. The input impedance will be parallel to these two ladder values.

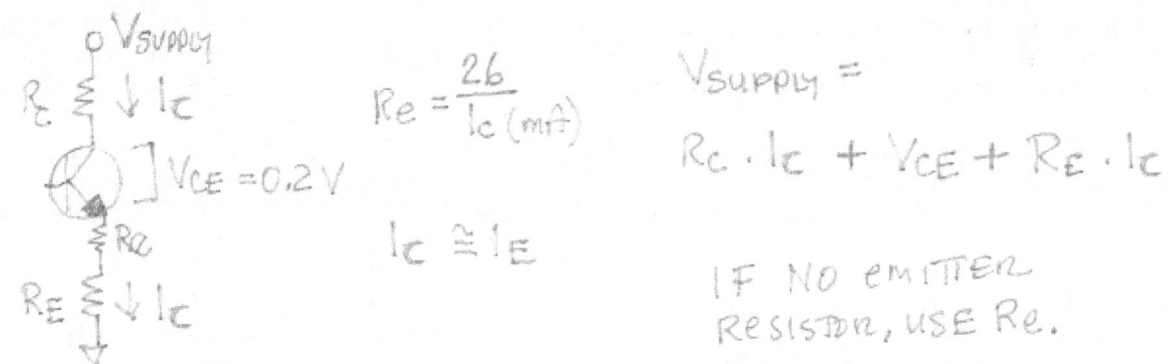

$$Re = \frac{26}{Ic \, (mA)}$$

$$Ic \cong IE$$

$$Vsupply =$$
$$Rc \cdot Ic + Vce + Re \cdot Ic$$

IF NO EMITTER
RESISTOR, USE Re.

Above is Kirchhoff's voltage law (KVL): the supply voltage equals the voltage drop across the collector resistor, Vce (0.2V for an MPSA18), the voltage drop across the internal resistance (re, consider if there is no emitter resistor), and the voltage drop across the emitter resistor.

$$Ic \cong IE$$

$$Re = \frac{26}{Ic \, (mA)}$$

$$Ic = hFE \cdot IB$$

$$Vsupply =$$
$$RB \cdot IB + VBE + Re \cdot Ic$$

IF NO EMITTER
RESISTOR, USE Re.

Above is Kirchhoff's voltage law (KVL): the supply voltage equals the voltage drop across the base resistor (Rb), Vbe (0.7V for an MPSA18), the voltage drop across the internal resistance (re, consider if there is no emitter resistor), and the voltage drop across the emitter resistor.

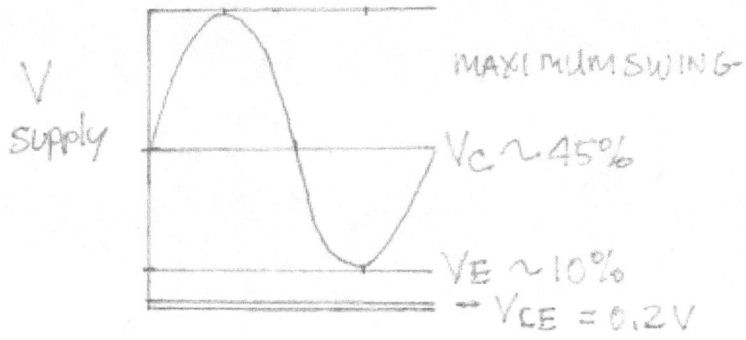

MAXIMUM SWING

Vc ~ 45%

VE ~ 10%
Vce = 0.2V

Above shows, graphically, how: 1) Vce takes up a tiny slice of voltage, 2) that 10% of the remaining voltage is put across the emitter for stability (negative feedback that prevents thermal runaway), and the 45% position of the collector so that it can swing maximally up and down. Other values for the percentages could be used; however, here, 10% of voltage on the emitter is used.

3. BJT BIASING SUMMARY

```
a. Pick supply voltage      supply = 1.5 V
b. Pick collector current Ic     Ic = 1 mA
c. Find base current Ib          Ib = Ic   / hFE
                                 Ib = 1 mA / 500
                                 Ib = 2uA
-------------------------------------------------------------------
A1. FIXED BIAS              Rb = (supply - Vbe) / Ib
    Find base resistor Rb   Rb = (1.5     - 0.7) / 2uA
                            Rb = 400k ohms


Note: With an emitter resistor, subtract its voltage drop.
For example: using the 0.150V Ve below:
                            Rb = (supply - Vbe - Ve   ) / Ib
                            Rb = (1.5     - 0.7 - 0.150) / 2uA
                            Rb = 325k ohms


A2. Find half voltage        V = (supply - Vce) / 2
                             V = (1.5     - 0.2) / 2
                             V = 0.650 V


A3. Find collector resistor Rc   Rc = Vc / Ic
                                 Rc = 0.650 / .001 A
                                 Rc = 650 ohms
-------------------------------------------------------------------
B1. LADDER BIAS             Ve = supply * 0.10
    Pick 10% supply for Ve  Ve = 1.5     * 0.10
                            Ve = 0.150 V
B2. Find emitter resistor Re   Re = Ve   / Ie
    Ie = Ic                    Re = .150 / .001 A
                               Re = 150 ohms
B3. Collector resistor Rc      Rc = 4.5 x Re (45% of supply)
    Half of 90% = 45%          Rc = 675 ohms
B4. Use emitter bypass cap     RF = 0.010 uF
                               audio = 10 uF (or more)
B5. Find base voltage Vb       Vb = Ve     + Vbe
                               Vb = 0.150 + 0.700
                               Vb = 0.850
B6. Pick bottom resistor R2    R2 = 100k ohms (to not load tank)
B7. Ladder equation            Vb = (supply * R2  ) / (R1 + R2)
    total = R1 + R2       0.850 = (1.5     * 100k) / (total  )
    R1 = top resistor     total = 176k
    R2 = bottom resistor     R1 = total - R2
                             R1 = 176k  - 100k
                             R1 = 76k ohms (top resistor)
```

The ladder should supply ~10 times the current needed by the base. By Ohm's law 1.5V/100k = 15 uA. We need 2 uA. Biasing for high collector current requires smaller R1 and R2 values. Note that the input impedance, at AC, is as if R1 and R2 are in parallel and attached to ground.

WORKING FOR HIGH VOLTAGE GAIN (BYPASS CAPACITOR)

```
a. Pick supply voltage       supply = 9.00 V
c. Vc = 45% of supply        Vc = 4.05 V
d. Ve = 10% of supply        Ve = 0.90 V
------------------------------------------------------------
c. Pick a current            Ic = 2 mA
d. Find Rc                   Rc = Vc/Ic
   Ohm's law                 Rc = 4.05 V / .002 A
                             Rc = 2025 ohms
------------------------------------------------------------
e. Find Re                   Ve = 0.9 V
   Ie = Ic                   Ie = 2 mA
   Ohm's law            Ve / Ie = 0.9 V / .002 A
                             Re = 450 ohms
------------------------------------------------------------
f. Note: unbypassed gain is only 2025/450 or 4.5
g. Note: bypassed gain is limited by output swing
h. Note: Vce of 0.2 not taken into account with a 9V supply
------------------------------------------------------------
i. Find Ib                   Ib = Ic / hFE
                             Ib = 2 mA / 500
                             Ib = 4 uA
j. Find Vb                   Vb = supply - Vce   - Ve
                             Vb = 9 V    - 0.7 V - 0.9 V
                             Vb = 7.4 V
k. Find Rb                   Rb = Vb / Ib
   Ohm's law                 Rb = 7.4 V / 4 uA
                             Rb = 1.85M ohm
------------------------------------------------------------
l. Note: the load, if other than Rc will effect gain.
m. Note: the load is as if in parallel to Rc.
------------------------------------------------------------
n. Xc = 1/(2*pi*f*C)         100uF = 1/(2*3.1415*300Hz*10uF)
   (much less than Re)       100uF = 5.31 ohms
------------------------------------------------------------
o. re internal resistance    re = Vt    / Ic
                             re = 26 mV /  2 mA
                             re = 13 ohms
p. Voltage gain = 2025/(13+5.31) = 110 gain
q. Changing Xc or the bypass capacitor will alter the gain.
   ex. 10 uF = 53.1 ohms, gain is 2025/(13+53.1) = 30
r. Note: at 1 uF the Xc is 531 ohms or greater than Re.
s. Note: higher frequencies will pass with less resistance.
```

4. JFET BIASING

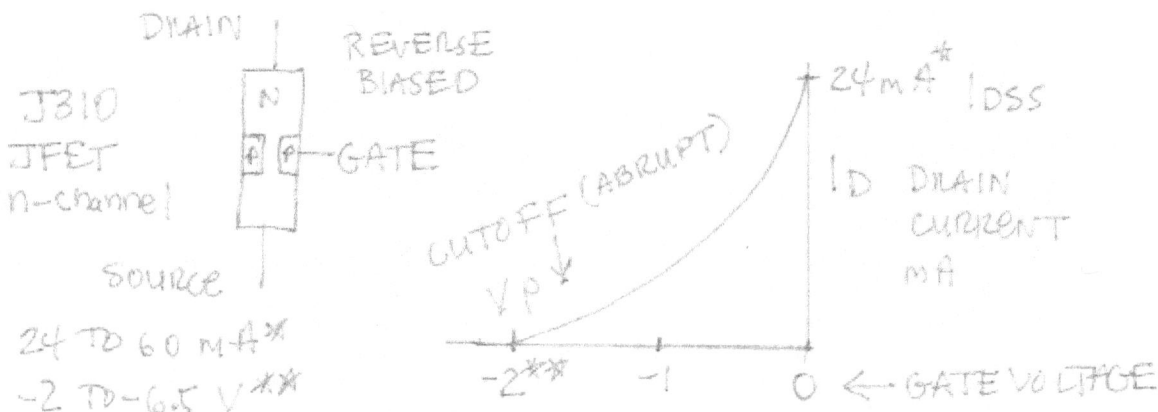

The curve above has an x-axis of gate-to-source voltage and y-axis of drain current (JFET device). At zero gate voltage there is a maximum current called Idss, that is between 24 to 60 mA (for the J310). As the gate becomes more negative, the depletion region expands, and decreases the drain current. At some point, Vp or Vgsoff, the drain current is zero. This is called cutoff. Sometimes JFETs are biased at cutoff, or class B operation. Only half the signal is amplified. Most of the time the JFET is biased to be between zero and maximum current. As a switch: at zero voltage, the switch is ON. This is using a drain load. And at minus ~2 volts the switch is OFF. This minus value for OFF can vary from -2 to -6.5 volts. The Achilles heel of the JFET is that its important parameters vary greatly even in a batch of similar stamped devices (ex. from one J310 to another).

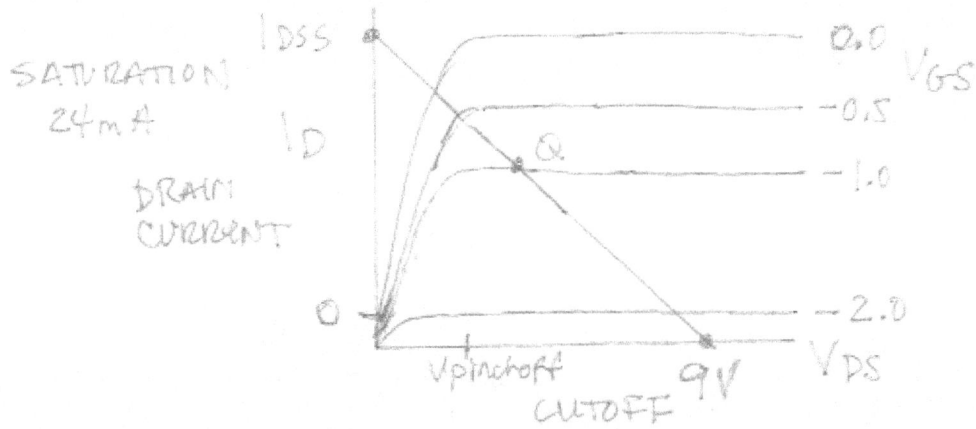

The graph above has an x-axis of drain-to-source voltage (Vds) and a y-axis of drain current (Id). The four curves show a decrease in current as the gate-to-source voltage (Vgs) drops from zero volts, to negative 0.5 volts, to negative 1.0 volts, and finally to negative 2.0 volts (no current). A JFET needs a higher amount of voltage to operate outside its resistor-like (ohmic) region. This minimal amount of voltage means that for a JFET to use a single battery, it must be a 9 Volt battery. Notice the load line with the Q point (representing minus one volt gate-to-source voltage). The line goes from 9 Volts or the supply voltage to maximum current or Idss. Below is the relationship between gate-to-source voltage (Vgs) and drain current (Id). These four points can be used to plot a curve of gate-to-source voltage versus drain current (for load line analysis). A resistor is a line that starts at "0,0" and goes to any voltage divided by current equal to its value.

Vgs		Id	
0.0	Vp	1.00	Idss
0.3	Vp	0.50	Idss
0.5	Vp	0.25	Idss
1.0	Vp	0.00	Idss

+9V

PINCHOFF
VOLTAGE

1M

+9V

100Ω

$\dfrac{voltage}{100} = Amps$

Idss

JFETS VARY

The gate-to-source voltage needed to cutoff the JFET (Vpinchoff) can be determined by attaching the drain to a + 9 Volt supply, attaching the gate to ground, and attaching the source to a 1M ohm resistor to ground (see above, left). The voltage at the source, with respect to ground, is the pinchoff voltage (Vgsoff). The Idss current or maximum current for the JFET can be determined by attaching the drain to a 100 ohm resistor hooked to a + 9 Volt supply, attaching the gate to ground, and attaching the source to ground (see above, right). The voltage across the 100 ohm resistor multiplied by 10 is the idss current in mA (or voltage divided by 100 is current in amps).

+V

AC

OUT

1M

Rs

COMMON
SOURCE
Rs SETS BIAS

+V

OUT

1M

Rs

COMMON
DRAIN
Rs SETS BIAS

+V

OUT

1M

IN

Rs

COMMON
GATE
Rs SETS BIAS

The easy way to bias a JFET is to have the AC source entering a small capacitor and then the gate. The gate is attached via a 1M ohm resistor to ground. This resistor keeps the gate at ground potential. This biasing is called automatic or self-biasing. Self-biasing uses a source resistor to raise the gate-to-source voltage potential. This also works for tanks where there is a direct DC path to ground (ex. through the tank inductor): the gate capacitor and resistor can be omitted. The gate voltage will be the same **absolute value** as at the top of the source resistor. This voltage can be measured. A potentiometer can be used as a source resistor and adjusted until the desired voltage (example: 1/3 of Vpinchoff) is obtained. The potentiometer can be measured and replaced with a standard resistor. Current can also be measured and set so half of Idss is obtained. A bypass capacitor around the source resistor (at radio or audio) is used to increase the voltage gain.

Vgsoff (Vpinchoff) can be determined (see procedure above). Set the gate-to-source (Vgs) voltage equal to Vgsoff divided by 3.4 to obtain a "voltage". Set drain current (Id) equal to Idss divided by 2 to obtain a "current". Divide the voltage by the current to get a source resistor (Rs) value. To calculate the drain resistor (Rd) take the the supply divided by 2 to obtain a voltage and divide by the current above (source and drain current are equal). These values can be measured.

If the supply is 9 Volts and the goal is 5.5 Volts (2V ohmic region) at the drain. Drain voltage equals supply minus drain current times drain resistance. Therefore 5.5 Volts equals 9 Volts minus 1 mA times drain resistance. Resistance is 3.5k ohms. These voltages can, again, be measured.

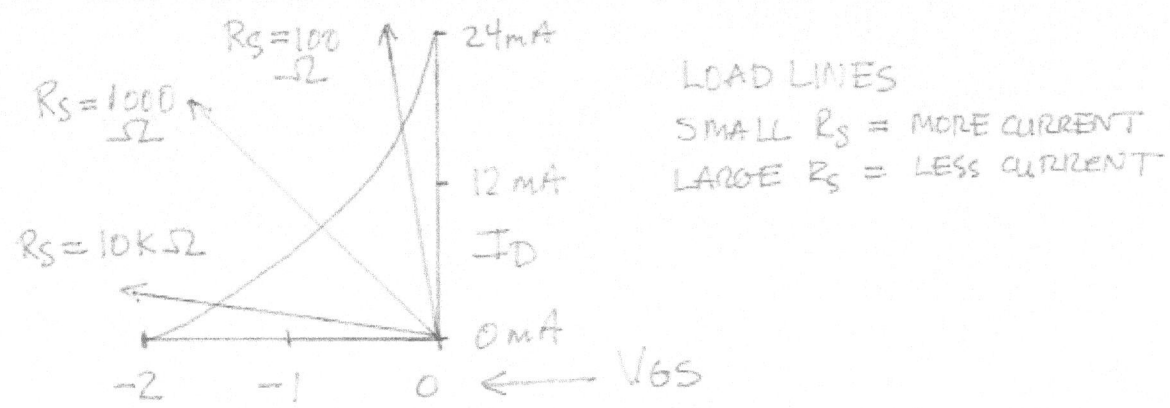

Above shows (NOT to scale) the effects of using different source resistor (Rs) values. Small source resistors increase current; while large source resistors decrease current. No source resistor would mean the Idss, here 24 mA of DC, is flowing in the circuit. A very large resistor would mean the Id is "near" zero and the voltage is near cutoff (this is why the circuit mentioned above uses a 1M ohm resistor to measure Vgsoff). Very low or very high source resistor values can cause class B operation. A good bias point is to use about one-third of Vgsoff or Vgsoff divided by 3.4. If a trimmer variable resistor is used in the source leg, it can be set until this value shows up above the source resistor. This is no-work biasing. Measure Vgsoff, divide by 3.4, and set it with a pot.

5. JFET BIASING SUMMARY

```
a. Pick supply voltage      supply =  9 V
b. Measure Vgsoff           Vgsoff = -4 V    or TYP data sheet value
c. Measure Idss               Idss = 10 mA   or TYP data sheet value
------------------------------------------------------------------
d. Pick half current    Idss * 0.49 = 4.9 mA    maximum current swing
e. See chart below    Vgsoff * 0.30 = 1.2 V
f. Source resistor      1.2V/.0049A = 245 ohms   Ohm's law
------------------------------------------------------------------
g. Drain resistor        half supply = 4.5 V
                          Idds * 0.49 = 4.9 mA
   Apply Ohm's law:      4.5V/.0049A = 918 ohms

h. For other ratios of Idss or Vgsoff use the chart below:
```

Idds * X	Vgssoff * X		Idds * X	Vgssoff * X
0.90	0.05		0.20	0.55
0.81	0.10		0.16	0.60
0.72	0.15		0.12	0.65
0.64	0.20		0.09	0.70
0.56	0.25		0.06	0.75
0.49	0.30		0.04	0.80
0.42	0.35		0.02	0.85
0.36	0.40		0.01	0.90
0.30	0.45		0.00	0.95
0.25	0.50		0.00	1.00

Ex. Pick Vgs as half (0.50) of Vgsoff. Idds will be a quarter (0.25).

```
WORKING FROM VOLTAGE GAIN (Rd)

a. Pick supply voltage        supply =  9 V
b. Measure Vgsoff             Vgsoff = -4 V   or TYP data sheet value
c. Measure Idss                 Idss = 10 mA  or TYP data sheet value
d. Pick a voltage gain         Vgain = 10
-----------------------------------------------------------------
e. Find Rd                     Vgain = gm     * Rd
   gm in S or mho "10mS"          10 = 0.010 * Rd
                                  Rd = 1000 ohms
f. Half supply at Vd              Vd = 4.5 Volts
g. Find Id                        Id =  Vd / Rd
   Ohm's law                      Id = 4.5 V / 1000 ohms
   4.5 mA                         Id = .0045 A
-----------------------------------------------------------------
h. Ratio Id/Idds             Id/Idds =  0.45
I. Chart Vgs/Vgsoff       Vgs/Vgsoff =  0.33
J. Calculate Vgs                 Vgs =  0.33 * -4
                                 Vgs =  -1.32
k. Is = Id                        Is = .0045 A
l. Find Rs                        Rs =   Vs / Is
   Ohm's law                      Rs = 1.32 / .0045
                                  Rs = 293 ohms




WORKING FROM VOLTAGE GAIN (Rd)

a. Pick supply voltage        supply = 18 V
b. Measure Vgsoff             Vgsoff = -6 V   or TYP data sheet value
c. Measure Idss                 Idss = 20 mA  or TYP data sheet value
d. Pick a voltage gain         Vgain = 30
-----------------------------------------------------------------
e. Find Rd                     Vgain = gm     * Rd
   gm in S or mho "20mS"          30 = 0.020 * Rd
                                  Rd = 1500 ohms
f. Half supply at Vd              Vd = 9 Volts
g. Find Id                        Id =  Vd / Rd
   Ohm's law                      Id = 9 V / 1500 ohms
                                  Id = 0.006 A
-----------------------------------------------------------------
h. Ratio Id/Idds             Id/Idds =  0.3
I. Chart Vgs/Vgsoff       Vgs/Vgsoff =  0.45
J. Calculate Vgs                 Vgs =  0.45 * -6
                                 Vgs =  -2.7
k. Is = Id                        Is = .006 A
l. Find Rs                        Rs =  Vs / Is
   Ohm's law                      Rs = 2.7 / .006
                                  Rs = 450 ohms
```

6. VACUUM TUBE BIASING

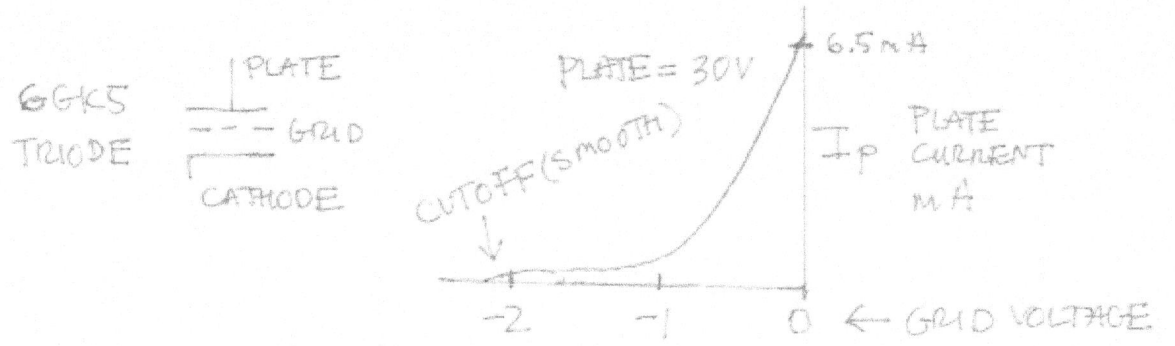

The curve above has an x-axis of grid voltage and y-axis of plate current (Ip). At zero grid voltage there is a maximum plate current: about 6.5 mA at 30 Volts of plate. As the grid becomes more negative, electrons are repelled, and the plate current decreases. At some point the plate current is zero: ~ minus 2.2 Volts. This is called cutoff. Sometimes tubes are biased at cutoff, which is class B operation. Only half the signal is amplified. Most of the time the tube is biased to be between no current and maximum current. As a switch at zero voltage the switch is ON. And at minus 2.2 volts the switch is OFF. This is using a plate load. The curve looks a lot like a JFET's.

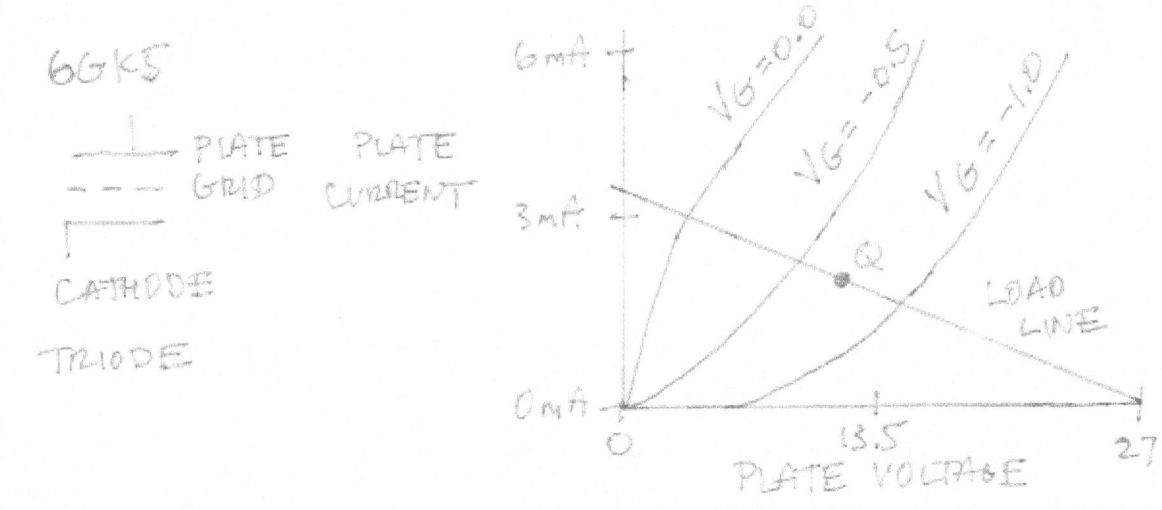

The graph above has an x-axis of plate voltage (Vp) and a y-axis of plate current (Ip). The three curves show a decrease in current as the grid-to-cathode voltage (Vg) drops from zero volts, to negative 0.5 volts, and finally to negative 1.0 volts. Notice the load line with the Q point (quiescent point). The load line goes from 27 Volts or supply voltage to maximum plate current.

Grids tend to become more negative. When electrons leave the cathode, flying in a vacuum, towards the positive plate voltage, some strike the grid. If the grid is not directly attached to ground (through a tank's inductor) or via a 1M or 10M ohm resistor (smaller resistors will drop the input impedance), the grid will collect electrons. This means that the grid will become more negative and eventually shut the plate current down. When the grid is high enough negative to shut the current down to zero in the plate, this is called cutoff. Cutoff is as if there is an open (not connected circuit). For any positive grid voltage, down to 0 volts (maximum plate flow), the tube is in saturation. Saturation is as if the device is a wire (closed circuit). For the graph, current will be the supply voltage divided by the load resistance. The input swings along the load line. For example, the Q point is about Vg = -0.75 volts. The input can swing up to -0.75 Volts or down to -0.75 Volts. If the signal is too great it will clip: the tube will either go into saturation or into cutoff.

To create triode curves: attach the cathode to ground, attach the plate to an ammeter then to the supply. And attach a variable negative supply (attach a +9 Volt battery with plus to ground, making minus at minus 9 Volts, relative) through a 10k-ohm resistor to the gate. Plotting on the x-axis gate voltage (Vg) and the y-axis plate current (Ip) will give lines with a slope of the gm or transconductance. Plotting on the x-axis plate voltage (Vp) and the y-axis plate current (Ip) will give lines with a slope of plate resistance (rp). Note that pentodes have much higher plate resistance. And that tube mu or amplification is gm (transconductance) times rp (plate resistance).

One way to bias a tube is to have the AC source enter a small capacitor and then the gate. The gate is attached via a 1M ohm resistor to ground. This resistor drains accumulated electrons off the gate. Electrons flying from the cathode to anode hit and accumulate on the grid. Without the 1M ohm resistor, this accumulation would lead to a progressively higher negative DC voltage and the grid shutting down plate current. The biasing itself is called automatic or self-biasing. Self-biasing uses a cathode resistor (Rk) to raise the grid voltage potential. This also works for tanks where there is a direct DC path to ground (through the tank's inductor). The voltage at the grid will be the same as at the top of the cathode resistor, only negative. This voltage can easily be measured. A potentiometer can be used as a source resistor and adjusted until the desired voltage (about 33% to 50% of cutoff) is obtained. The potentiometer can be measured and replaced with a standard resistor. Or a trimmer pot can be used. A bypass capacitor (at radio or audio) can be used to increase the voltage gain. It can be thought of as setting the cathode, at RF or AF, to ground.

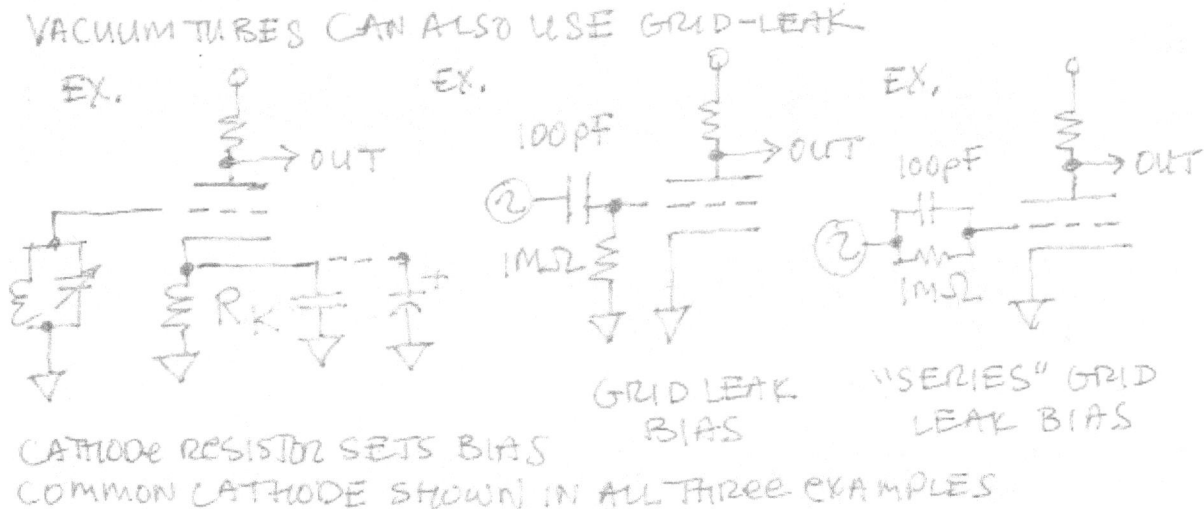

VACUUM TUBES CAN ALSO USE GRID-LEAK

EX.   EX.   EX.

100 pF   →OUT   100pF   →OUT

1MΩ   1MΩ

GRID LEAK BIAS   "SERIES" GRID LEAK BIAS

CATHODE RESISTOR SETS BIAS
COMMON CATHODE SHOWN IN ALL THREE EXAMPLES

The easiest way to bias a tube is called grid leak biasing (above center and right). The grid leak maintains a stable negative voltage on the grid. The RF source enters a 100 pF capacitor that is attached to the grid. The grid is attached through a 1M ohm resistor to ground (above center). A variation of grid leak biasing is to have the RF source enter both the 100 pF capacitor and 1M ohm resistor (in parallel) and then enter the grid (above right). Grid-leak bias works well on filament tubes. These are tubes that do not have a separate cathode hookup: ex. 6418 or 6612. The 6GK5 is an indirectly-heated tube. This mean it has a separate cathode connection for a cathode resistor.

The plate load is chosen to increase voltage gain. Voltage gain equals mu times Rload divided by plate resistance plus Rload. This basically means that load resistance should be as high as possible, but no higher than 10 times the plate resistance. The plate resistance at low plate voltages on a triode will be fairly low (ex. 5000 ohms). The plate resistance of a pentode will be fairly high (ex. 10M ohms). Plate resistance often drops dramatically as the gate voltage approaches zero volts. This means it is highest when biased near cutoff. Where gm is lowest.

7. JFET AND VACUUM TUBE DIODE BIASING

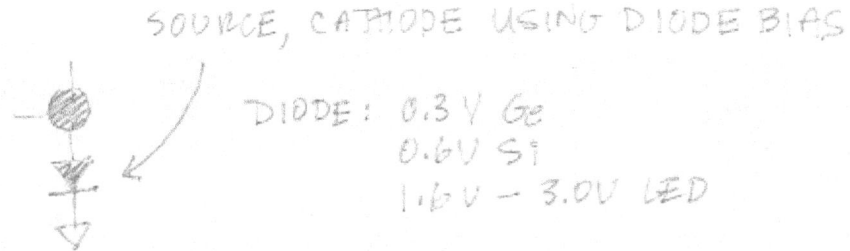

SOURCE, CATHODE USING DIODE BIAS

DIODE: 0.3 V Ge
0.6V Si
1.6V - 3.0V LED

A diode from source to ground (JFET) or cathode to ground (triode or pentode) can be used to establish an exact value of voltage on the gate (JFET) or grid (vacuum tube). A germanium diode will set gate (JFET) or grid (tube) voltage to ~0.3V. A silicon diode will set them to ~0.7V. A 1N4148 (silicon, small signal, fast switching diode) will set them to ~1.0V. And various LED (light emitting diodes) will set them to ~1.6V to 3.0V. The gate (JFET) or grid (vacuum tube) will be set at ground potential for DC. This can be done via attaching a 1M ohm resistor between the gate or the grid and ground or via the coil of a tank. Half the RF present on the gate or grid will be trapped above the diode. This can radiate or can be used (see designs) as a source for oscillator feedback.

8. DIODE VOLTAGE DROPS

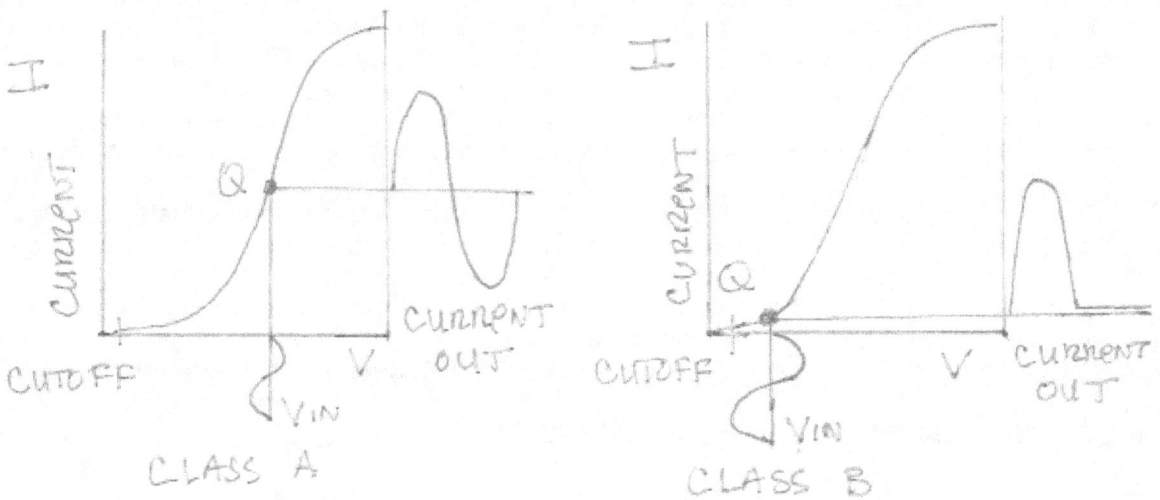

A diode will drop the voltage available to the rest of the circuit. The current shown is the maximum each diode can handle safely. Solving: 10 Volts minus 0.7 Volts leaves 9.3 Volts. Current is 0.10 amp. By Ohm's Law, V=I*R; therefore, R1 equals 930 ohms. Solving: 5 Volts minus 2.3 Volts leaves 2.7 Volts. Current is 0.005 amp. By Ohm's Law V=I*R; therefore, R2 equals 540 ohms. These resistors can be higher and this will just mean less current and lower LED output. However if the resistor is lower than those calculated the diode may be damaged. It should also be noted that resistors only come in certain values, typically the closest is chosen. In this circumstance, pick the next highest standard resistor value. Meaning: R1 would be 1.0k ohms and R2 would be 560 ohms.

9. CLASS OF OPERATION

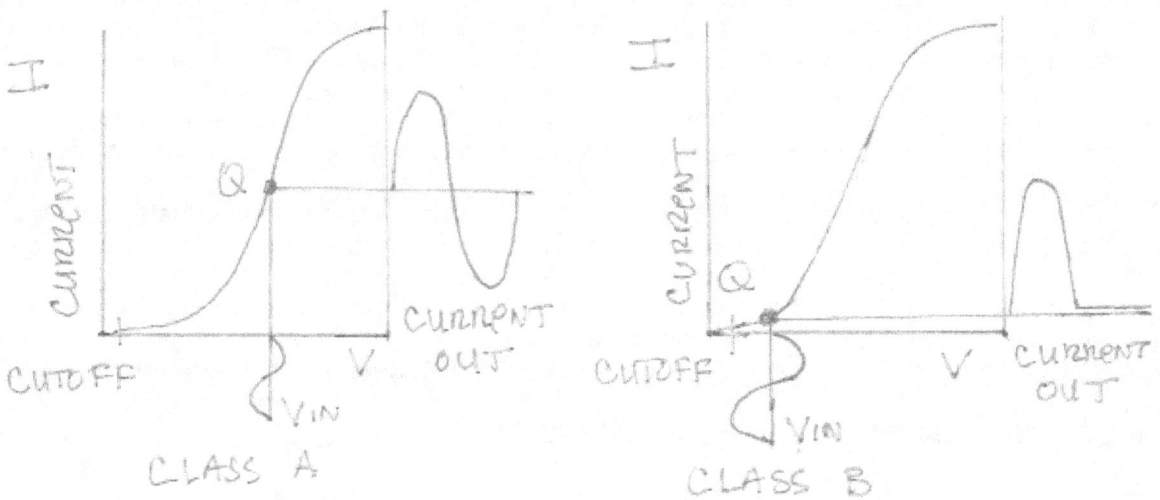

The left graphic shows class A amplification: the Q point is in the middle of the JFET transfer curve. Note that a small voltage in (seen above "Class A") leads to a large current out (right). The right graphic shows class B amplification: the Q point is near cutoff. Note that a small positive voltage swing in (see above "Class B") results in a large current out (right); however, the negative voltage swing results in below cutoff or no current flowing. This is the flat line on the right after the large voltage swing upward. Class A and class B operation can be setup in a BJT, as shown below. The Q point labeled "A" is for class A operation. The Q point labeled "B" is for class B operation. The AC signals shown on the graph are the inputs: the output voltage is downward (see arrow) and the output current is shown to the left of the Q point. At point Q$_B$, the transistor is biased at cutoff.

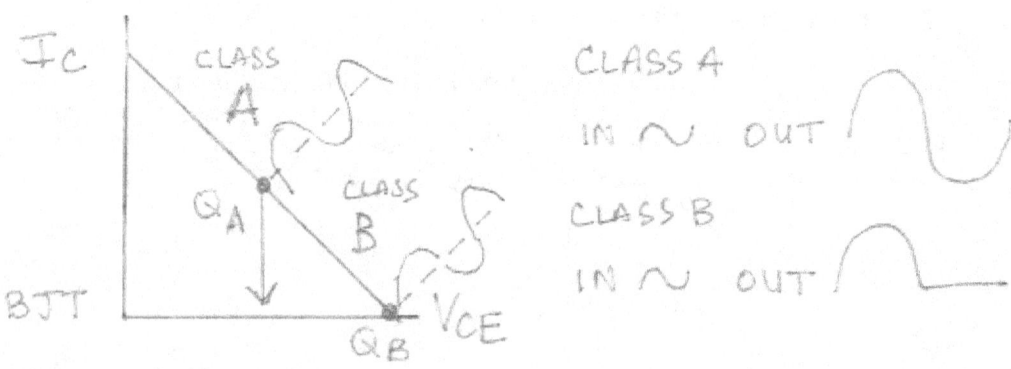

1.9 Amplifier Data Appendix
THE RADIO BUILDING HOBBY
©2015

1. OVERVIEW OF MODES OF OPERATION

	#1	#2	#3
Common (BJT)	emitter	collector	base
Common (JFET)	source	drain	gate
Common (triode)	cathode	plate	grid
Current (I) gain	high	high	1
Voltage (V) gain**	high	1	high
Power gain	high	medium	low
Inverted	yes	no	no
Input	#3	#3	#1
Output	#2	#1	#2
Z in	high*	high	low
Z out	medium	low	high

Where (*) is medium for a BJT and (**) tubes have low voltage gain. Tubes designs often make up for this lack of voltage gain by using a 1:10 transformer as the plate load.

```
Resistors — limit current
     Resistances over 250k ohms starts acting like it is "not connected".
Capacitors — store electrons
     Xc = 1/(2*pi*fC) //current leads voltage by 90 degrees
     where pi = 3.14159, f = frequency, and C is capacitance in Farads.
Inductors — oppose current change
     Xl = 2*pi*fL      //current lags voltage by 90 degrees Q = Xl/R
     where pi = 3.14159, f = frequency, and L is inductance in Henries.
Transformer — mutual inductance
     2 turns: 4 turns = double voltage, half current
     6 turns: 3 turns = half voltage, double current
Tank frequency = 1/(2*pi*squareroot(LC))
     pi = 3.14159
     L is inductance in Henries.
     C is capacitance in Farads.
     Increase L or C to reduce frequency.
Ohm's law: V = I * R   //voltage = current times resistance
Power = V * I          //power = voltage times current
Voltage, in series, is dumped into the largest load.
Current, in series, is the same through each load.
Therefore, power, in series, is dumped into the largest load.
Z is impedance (in ohms)
     Low Z = incoming RF
     High Z = tank
     Want under a 10 times mismatch in Z for best power transfer.
```

1.10 MPSA18 Data Appendix
THE RADIO BUILDING HOBBY
©2015

1. BIPOLAR JUCNTION TRANSISTOR MODES: NPN

BJT (NPN)	common emitter	common collector	common base
Current (I) gain	high	high	one
Current (I) gain	hFE	hFE	1
MPSA18	500	500	1
Voltage (V) gain	high	one	high
w/bypass cap	-gm x Rc	1	gm x Rc
no bypass cap	-Rc/Re		
Inverted	yes	no	no
MPSA18	-1520	1	1520
Input impedance	base	base	emitter
Z in	medium	high	low
Z in	hFE x Re	hFE x Re	1/gm
Z in (ohms)	13k	500k+	26
Output impedance	collector	emitter	collector
Z out	medium	low	high
Z out	Rc	1/gm	>> Rc
Z out (ohms)	40k	26	40k

Where hFE is beta or current gain; gm is transconductance, Rc is collector resistance, and Re is emitter resistance (internal and external). This describes the three modes of operation.

2. BJT COLLECTOR CURRENT VERSUS INTERNAL EMITTER RESISTANCE

Ic (mA)	gm (S)	re or 1/gm (ohm)
1	0.038	26.00
2	0.077	13.00
5	0.192	5.20
10	0.385	2.60
25	0.962	1.04
100	3.846	0.26

Where Ic is collector current, gm is transconductance in Semens (S), and re is the internal emitter resistance. The transistor is the MPSA18. The re component gives rise to unstable voltage gain without an external emitter resistor. This can cause thermal runaway: wherein higher current causes lower re which in turn, causes higher current. Consider common emitter (emitter to ground) mode with a 10k ohm collector load. Voltage gain is not infinite or 10,000 but 384 due to the emitter resistance of 26 ohms at 1 mA. Voltage gain will be higher at higher collector currents.

3. BJT BIASING RESISTOR VERSUS COLLECTOR CURRENT

Rb (ohm)	Ib (uA)	Ic (mA)
3.7k	190.48	95.24
10k	76.19	38.10
47k	16.84	8.42
110k	7.24	3.62
220k	3.63	1.81
470k	1.70	0.85
1M	0.80	0.40

Where Rb is the base resistor, Ib is base current, and Ic is collector current. This is for the MPSA18. Parameters: hFE is 500, supply is 1.5 Volts, and 1-ohm emitter and collector loads.

4. BJT OVERVIEW AND EQUATIONS

A BJT transistor is a current controlled device. Common base has very high output impedance (parallel to load). Common collector is called an emitter follower. Common emitter has the highest power gain. A BJT has very high transconductance but low input impedance. The base to emitter junction is a forward biased PN junction. Meaning: it needs 0.7 Volts to operate (see data sheet Vbe value). The base to emitter has a 0.7 Volt voltage drop. Current gain is beta or hFE.

```
re = 1/gm = 26 mV/Ie = 0.26V/Ie    //re is emitter resistance
re varies with emitter current (Ie) and is temperature unstable.
Total Re includes 1/gm and the emitter resistor.
Total Re includes 1/gm and bypass cap's capacitive reactance at RF or audio.

Ic = hFE * Ib                       //Ic is collector current
Ib = (supply - 0.7) /  Rb           //fixed or base bias
Ib = (supply - 0.7) / (Rb + hFE x Rc)    //collector feedback bias
Ib = (supply - 0.7) / (Rb + hFE x Re)    //fixed bias with emitter resistor
Where supply is X volts, Ib is base current, and Rb is the base resistor.

Vb = (supply * R2) / (R1 + R2)    //resistor ladder biasing: R1 top, R2 bottom
Vb = Ie * Re = Ic * Re            //Ie = Ic (approximately)
Rb = (R1 x R2) / (R1 + R2)        //can set R1 = R2, make >100k

Vc = supply — Ic x Rc    //Vc is collector voltage
Ie = Ic                  //approximate, Ie is emitter current
Ve = Vb  - 0.7           //Vbe = 0.6 to 0.7 (Si) and 0.2 to 0.3 (Ge)
Ve = Ic x Re             //Ve is emitter voltage
Vce = Vbe + Vcb          //Used for collector feedback bias (KCL)

Vc for CE/CB = 0.45 x supply    //largest output swing + 0.10 on emitter
Ve for CC   = 0.45 x supply    //largest output swing + 0.10 on emitter

1) Saturation: Ic = supply/(Rc + Re)    //closed switch or maximum current
2) Cutoff:    Vce = supply              //open switch or no current
3) Active region: linear relationship between Ib and Ic

Class A is 25.0% efficient.
Class B is 78.5% efficient (works near cutoff).
```

Rmax mA	1.5 Volts	3.0 Volts	9.0 Volts	18 Volts
0.25	6,000	12,000	36,000	72,000
0.5	3,000	6,000	18,000	36,000
1	1,500	3,000	9,000	18,000
2	750	1,500	4,500	9,000
5	300	600	1,800	3,600
10	150	300	900	1,800
20	75	150	450	900
50	30	60	180	360

10% Re mA	1.5 Volts	3.0 Volts	9.0 Volts	18 Volts
0.25	520	1,120	3,520	7,120
0.5	260	560	1,760	3,560
1	130	280	880	1,780
2	65	140	440	890
5	26	56	176	356
10	13	28	88	178
20	7	14	44	89
50	3	6	18	36

45% Rc mA	1.5 Volts	3.0 Volts	9.0 Volts	18 Volts
0.25	2,340	5,040	15,840	32,040
0.5	1,170	2,520	7,920	16,020
1	585	1,260	3,960	8,010
2	293	630	1,980	4,005
5	117	252	792	1,602
10	59	126	396	801
20	29	63	198	401
50	12	25	79	160

Rbase mA	1.5 Volts	3.0 Volts	9.0 Volts	18 Volts
0.25	1,300,000	4,000,000	14,800,000	31,000,000
0.5	650,000	2,000,000	7,400,000	15,500,000
1	325,000	1,000,000	3,700,000	7,750,000
2	162,500	500,000	1,850,000	3,875,000
5	65,000	200,000	740,000	1,550,000
10	32,500	100,000	370,000	775,000
20	16,250	50,000	185,000	387,500
50	6,500	20,000	74,000	155,000

All values in the chart are in ohms. The rows above show different supply voltages: 1.5, 3, 9, and 18 Volts. The columns to the left show collector current in mA. Rmax is the maximum resistance that allows the current under it: via Ohm's law. Re 10% is the resistance for 10% of supply drop across the emitter. Rc 45% is the resistance for 45% of supply drop across the collector. Rbase is the base resistance needed to provide the collector current in mA under it.

ohm load	250	500	1000
Rbase	66,667	133,333	266,667

ohm load	2500	5000	10000
Rbase	666,667	1,333,333	2,666,667

Above is a list of base resistors (Rbase) needed to cause saturation for a given ohm collector load. The conditions are: supply = 1.5V, hFE = 500, and Vbe = 0.7V. **Supply is 1.5V.**

ohm load	250	500	1000
Rbase	115,278	230,556	461,111

ohm load	2500	5000	10000
Rbase	1,152,777	2,305,556	4,611,111

Above is a list of base resistors (Rbase) needed to cause saturation for a given ohm collector load. The conditions are: supply = 9.0V, hFE = 500, and Vbe = 0.7V. **Supply is 9.0V.**

5. BJT DARLINGTON PAIR TRANSISTOR BIASING

In a Darlington transistor's "base" to "emitter" circuit, there are two forward voltage drops of 0.7 Volts or 1.4 Volts total. A single 1.5 Volt battery would be a poor choice. Using two batteries in series, or 3.0 Volts, the voltage available, after the voltage drops, is 1.6 Volts. For a current of 1 mA at the collector, a base current of 0.1 uA is needed. This is 1 mA divided by the hFE of 10,000. Using Ohm's law 1.6 Volt divided by 0.1 uA is 16M ohms. This is the resistor needed from positive power to the base, for biasing. This could be made using several, smaller resistors in series. Using a 9 Volt battery power supply would lead to needing a bias resistor in the range of 76M ohms! A Darlington transistor, such as the MPSA14, can be thought of as a single active device, similar to a pentode. Where a pentode vacuum tube adds more grids, a transistor adds more N and P regions. The device still has three terminals. However, the current gain is only about 20 times higher than the MPSA18's hFE of 500. Due to the use of regeneration, the MPSA18 will suffice for simple radios.

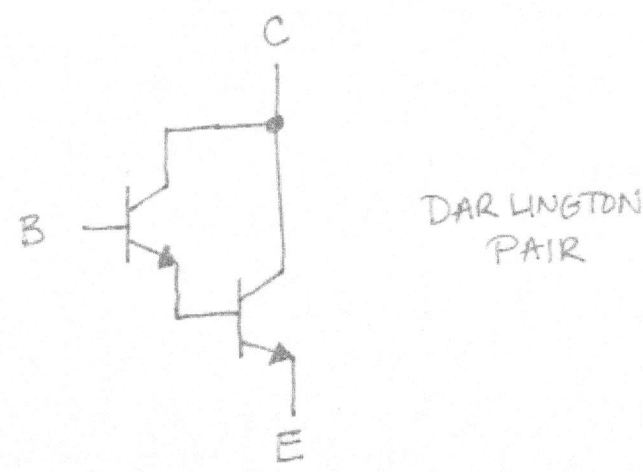

DARLINGTON PAIR

47

1.11 J310 Data Appendix
THE RADIO BUILDING HOBBY
©2015

1. JUNCTION FEILD EFFECT TRANSISTOR MODES: N-CHANNEL

JFET (N-channel)	common source	common drain	common gate
Current (I) gain	high	high	one
Current (I) gain	infinite	infinite	1
J310	1G	1G	1
Voltage (V) gain	high	one	high
w/bypass cap	-gm x Rd	1	gm x Rd
no bypass cap			
Inverted	yes	no	no
J310	-320	1	320
Input impedance	gate	gate	source
Z in	high	high	low
Z in	300M	300M	1/gm
Z in (ohms)	10M	10M	125
Output impedance	drain	source	drain
Z out	medium	low	high
Z out	Rd	1/gm	>> Rd
Z out (ohms)	40k	125	40k

Where gm is transconductance, Rd is drain resistance, and Rs is source resistance.

2. JFET VARIATION IN TRANSCONDUCTANCE

gm (S)	1/gm (ohm)
0.008	125
0.010	100
0.012	83
0.013	77
0.014	71
0.016	63
0.018	56

Where gm is transconductance: 0.008 to 0.018 S is the variation for a J310 JFET. This is also 8 mS to 18 mS, where S is Siemens. Notice the variation in common-gate input impedance (Zin) that depends on the gm of the individual device. Fortunately, this variation is only a factor of two.

3. JFET Rs AND Rd WITH VARIABLE Vgs, Vgsoff, AND Idss.

Rs	Vgs	Vgsoff	Idss	Rd
68	-1.0	-2	0.060	240
167	-1.0	-2	0.024	560
75	-1.0	-4	0.024	110
62	-1.0	-6	0.024	10
333	-2.0	-4	0.024	430
500	-3.0	-6	0.024	270
390	-1.0	-2	0.010	1500
820	-1.0	-2	0.005	3000
1300	-1.0	-2	0.003	4700
2000	-1.0	-2	0.002	6800
3900	-1.0	-2	0.001	15000
24000	-1.5	-2	0.001	62000
910	-0.5	-2	0.001	5600

Where Rs is source resistance, Vgs is gate-to-source voltage, Idss is drain current at zero bias, Vgsoff is the cutoff voltage, and Rd is drain resistance. The device is the J310, supply is 9 Volts, gate resistor is 1M ohm, and internal drain resistance is 4000 ohms. The variances of Vgsoff and Idss are typical for the J310. Notice the differences in Rs and Rd with varying JFET values.

4. JFET OVERVIEW AND EQUATIONS

A JFET transistor is a voltage controlled device. Common gate has very high output impedance (parallel to load). Common drain is called a source follower. Common source has the highest power gain. A JFET has very high input impedance but needs drain voltage to operate. Input impedance is 300M to 1G ohms, as it is a reverse biased PN junction. Input impedance is in parallel to the 1M+ gate resistor. Current gain is "infinite" due to slight current input. Grid-leak is not necessary with a JFET. Idss and Vp vary and can make it hard for others to duplicate a design.

```
Vneeded = Vp + Idss * Rload   //pinch-off = -Vgsoff
                              //Vneeded = 8V @ 1 mA w/1.5k load (example)
Vgs = -Id * Rs                //Vgs for self-bias
Vs  =  Id * Rs                //Vs or source voltage
Vd = supply - Id * Rd         //Vd or drain voltage
Is = Id                       //Is or source current
Ig = 0                        //gate current is zero
supply = Vds + Id * Rd + Id * Rs   //KVL

Vgs = Vg — Vs                     //Vgs or gate-to-source voltage
Vds = Vd — Vs                     //Vds or drain-to-source voltage
Id = Idss (1 — Vgs/Vgsoff)^2  //Id or drain current (see ratio chart below)
```

```
Bias: voltage source applying negative voltage via a large gate resistor.
Bias: current source in source leg.
Self-bias: a large gate resistor to ground.
          The source resistor creates gate voltage.
Self-bias: is sensitive to JFET parameter spread.
Bias: A diode in the source leg can set voltage at the gate.
Rs = (Vgsoff/3.333)/(Idss/2)   //Rs for half Idss current
Rd = (supply/2)/(Idds/2)       //Rd for half supply at Vd
```

5. GATE VOLTAGE VERSUS CURRENT RATIOS

Vgssoff * X	Idds * X	Gm * X	Idds * X	Vgssoff * X	Gm * X
0.00	1.0000	1.0000	0.00	1.0000	0.0000
0.02	0.9604	0.9800	0.02	0.8586	0.1414
0.04	0.9216	0.9600	0.04	0.8000	0.2000
0.06	0.8836	0.9400	0.06	0.7551	0.2449
0.08	0.8464	0.9200	0.08	0.7172	0.2828
0.10	0.8100	0.9000	0.10	0.6838	0.3162
0.12	0.7744	0.8800	0.12	0.6536	0.3464
0.14	0.7396	0.8600	0.14	0.6258	0.3742
0.16	0.7056	0.8400	0.16	0.6000	0.4000
0.18	0.6724	0.8200	0.18	0.5757	0.4243
0.20	0.6400	0.8000	0.20	0.5528	0.4472
0.22	0.6084	0.7800	0.22	0.5310	0.4690
0.24	0.5776	0.7600	0.24	0.5101	0.4899
0.26	0.5476	0.7400	0.26	0.4901	0.5099
0.28	0.5184	0.7200	0.28	0.4708	0.5292
0.30	0.4900	0.7000	0.30	0.4523	0.5477
0.32	0.4624	0.6800	0.32	0.4343	0.5657
0.34	0.4356	0.6600	0.34	0.4169	0.5831
0.36	0.4096	0.6400	0.36	0.4000	0.6000
0.38	0.3844	0.6200	0.38	0.3836	0.6164
0.40	0.3600	0.6000	0.40	0.3675	0.6325
0.42	0.3364	0.5800	0.42	0.3519	0.6481
0.44	0.3136	0.5600	0.44	0.3367	0.6633
0.46	0.2916	0.5400	0.46	0.3218	0.6782
0.48	0.2704	0.5200	0.48	0.3072	0.6928
0.50	0.2500	0.5000	0.50	0.2929	0.7071
0.52	0.2304	0.4800	0.52	0.2789	0.7211
0.54	0.2116	0.4600	0.54	0.2652	0.7348
0.56	0.1936	0.4400	0.56	0.2517	0.7483
0.58	0.1764	0.4200	0.58	0.2384	0.7616
0.60	0.1600	0.4000	0.60	0.2254	0.7746
0.62	0.1444	0.3800	0.62	0.2126	0.7874
0.64	0.1296	0.3600	0.64	0.2000	0.8000
0.66	0.1156	0.3400	0.66	0.1876	0.8124
0.68	0.1024	0.3200	0.68	0.1754	0.8246
0.70	0.0900	0.3000	0.70	0.1633	0.8367
0.72	0.0784	0.2800	0.72	0.1515	0.8485
0.74	0.0676	0.2600	0.74	0.1398	0.8602
0.76	0.0576	0.2400	0.76	0.1282	0.8718
0.78	0.0484	0.2200	0.78	0.1168	0.8832
0.80	0.0400	0.2000	0.80	0.1056	0.8944
0.82	0.0324	0.1800	0.82	0.0945	0.9055
0.84	0.0256	0.1600	0.84	0.0835	0.9165
0.86	0.0196	0.1400	0.86	0.0726	0.9274
0.88	0.0144	0.1200	0.88	0.0619	0.9381
0.90	0.0100	0.1000	0.90	0.0513	0.9487
0.92	0.0064	0.0800	0.92	0.0408	0.9592
0.94	0.0036	0.0600	0.94	0.0305	0.9695
0.96	0.0016	0.0400	0.96	0.0202	0.9798
0.98	0.0004	0.0200	0.98	0.0101	0.9899
1.00	0.0000	0.0000	1.00	0.0000	1.0000

6. J310 VARIABILITY

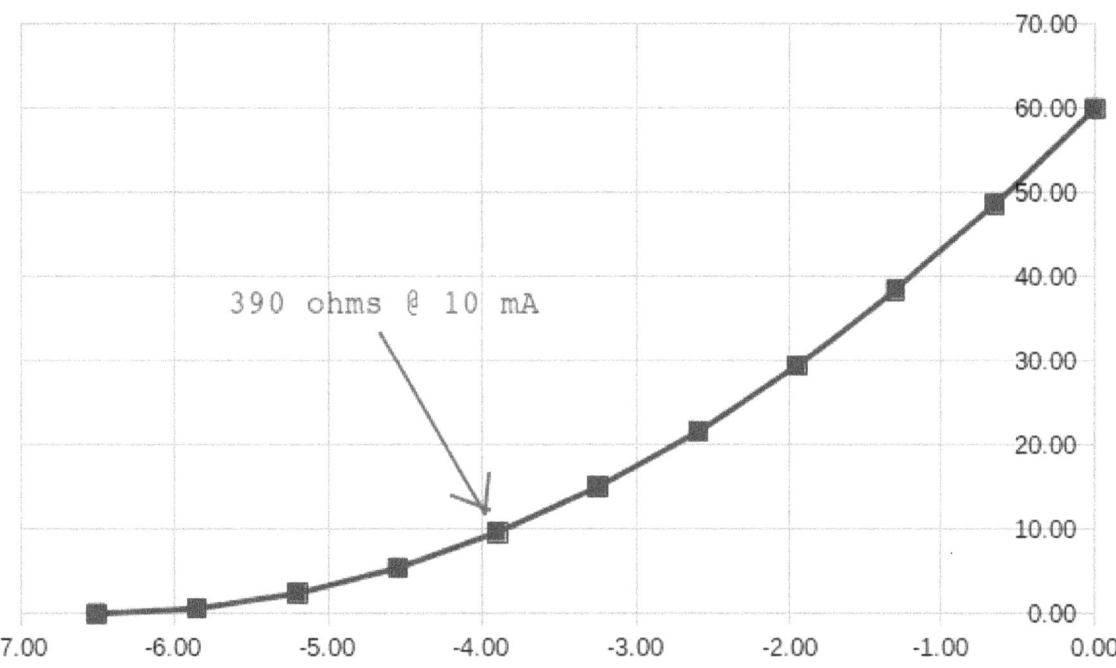

Above is the J310 using its maximum Vgsoff of -6.5V and highest current of 60 mA. To maintain a bias of 10 mA requires -3.90V gate-to-source. By Ohm's law this is 390 ohms.

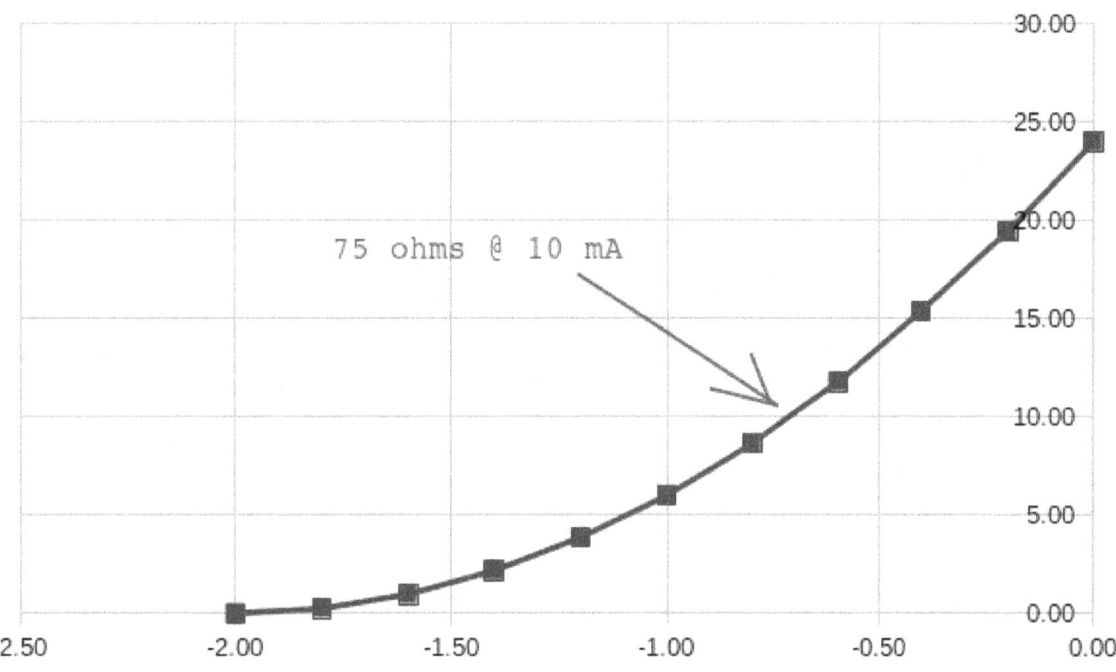

Above is the J310 using its minimum Vgsoff of -2.0V and lowest current of 24 mA. To maintain a bias at 10 mA requires -0.75V gate-to-source. By Ohm's law this is a 75 ohms.

The two graphs illustrate the difficulty in nailing down a source resistor value when using self-biasing. It is best for a design to use a trimmer resistor as the source resistor so that variations in individual devices can be accommodated. This was done in the JFET Hellenedyne design. In the Science Fair Radio and Heliosdyne the JFET resistors are large and the device is near cutoff. A J310 may need 8 or more Volts to operate; whereas, a BJT can operate on a single 1.5 Volt battery.

7. J310 BIASING GRAPHING

Vgs		Id	
0.0	Vp	1.00	Idss
0.3	Vp	0.50	Idss
0.5	Vp	0.25	Idss
1.0	Vp	0.00	Idss

Measurements were made on an individual J310 having a Vgssoff of -4 Volts and an Idss of 40 mA. These can be plotted on a graph using the four points from the chart above. The data comes out to be (Vgs, Id) of: (0.0 V, 40 mA), (-1.2 V, 20 mA), (-2.0 V, 10 mA), and (-4.0 V, 0 mA). From these four points a rough curve can be drawn or approximated using graphing software.

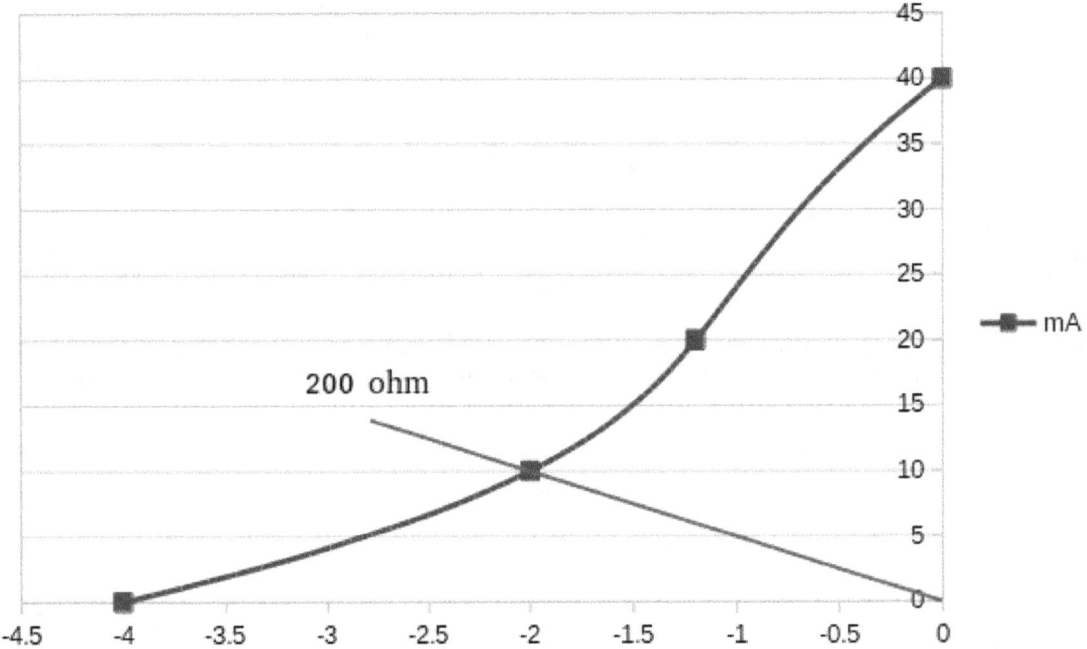

The source resistor needed to create a bias current of 10 mA at -2 Volts is 200 ohms. This is via Ohm's law: dividing 2 V by .010 A is 200 ohms. The graph is not needed to select this resistor. Any ratio of Vg to Vgsoff or ratio of Id to Idss can be looked up on the table in section "5" above and the other value calculated. What the graph shows nicely is how much more or less the gate voltage and drain current can change due to the size of the source resistor. By using a larger biasing source resistor, the swing of input and drain current are reduced. In radio, the signal may be as weak as 0.1 uV at 0.01 uA. A regenerative radio will increase voltage by about 8000 times. This brings the signal up to about 0.8 mV or greater. And crystal earphones are very sensitive.

The drain resistor can be chosen for gain (gm times drain resistance) or to place half the supply, ex. 4.5 Volts for a 9V battery on the drain. For 4.5 Volts to be present when the current is 10 mA or 0.010 A requires a resistance of 450 ohms. Voltage gain is approximately gm times resistance or 13,000 umhos or 13,000 uS or 13 mS times 450 or 5.85, which is not stellar but current gain is very high. This voltage gain is with a bypass resistor. The gain in dB is equal to 15.3 dB. To convert a voltage gain to dB: take the log of the voltage gain and then multiply the result by 20. To convert a dB to voltage gain: raise 10 to the power of the number in dB divided by 20.

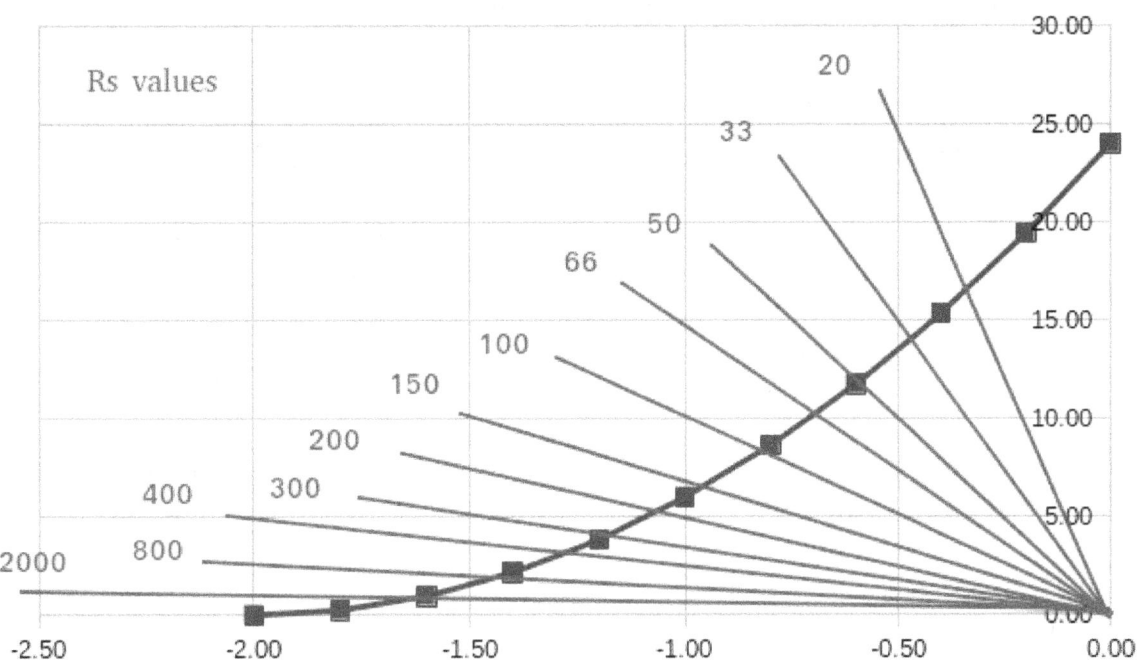

The above graphic shows the effect of altering the source resistor on a J310 JFET using its minimum specs for idss (24 mA) and Vgsoff (-2 V). A resistor is shown as a straight line coming from point 0,0 and then passing through all values of voltage divided by current that equal its resistance. For example, "20 ohms" passes through the point -0.50 V and .025 A (25 mA): .5 divided by .025 is 20. This single point was used to draw the line. What is readily apparent is that it takes at least 150 ohms to bring the steady-state current down to about 10 mA. Realize that in the past, a tube might require hundreds of volts to produce that much plate current. Also notice that at the highest bias source resistance value, 2k ohms, that there is still 1.25 mA of current flowing and a half volt (from -1.50 to -2.00 V) voltage that can swing on the input signal. Most radio signals are from about S3 or 0.8 uV to S9 or 50.2 uV. With maximum regenerative gain of 8,000 this becomes 0.8 mV to 50.2 mV. The JFET's 0.5 V swing represents 500 mV. There is still plenty of room for an S9 level signal when using a regenerative radio to its full potential.

In a JFET, the voltage drop across the source resistor and the difference it causes in Vgs (voltage gate-to-source becomes negative by the same amount) controls the current that flows. The gate is negative, relative to the source. This is called self-biasing. The drain resistor could be any value. It can be chosen for maximum voltage swing on the drain. Or it can be chosen to set a gain through the formula: voltage gain equals gm times Rd. The bypass resistor attached to the source can be though of as bringing that point to being "ground" for AC: radio, audio, or both.

The JFET can be operated at saturation by using no source resistor and the gate attached to ground. This will consume a lot of current: the amount being equal to the Idss specification. The JFET can be operated at cutoff by using a high-value source resistor. To completely shut down the JFET would take an infinitely large resistor. However, for practical purposes, any large resistor (ex. 100k) will suffice as being "cutoff". Operating in this mode will consume very little current: about zero. In either mode, a swing in only one direction will cause a decrease (when at saturation) or an increase (when at cutoff) in drain current. This can be used as a detector. The detector will have voltage and current gain at the drain. And the detector will have current gain at the source.

1.12 Vacuum Tube Data Appendix
THE RADIO BUILDING HOBBY
©2015

1. VACUUM TUBE MODES: TRIODE

TUBE (triode)	common cathode	common plate	common grid
Current (I) gain	high	high	one
Current (I) gain estimate	estimate	estimate	1
6GK5	1G	1G	1
Voltage (V) gain	low	one	low
w/bypass cap	-gm x Rp	1	gm x Rp
no bypass cap	less		
Inverted	yes	no	no
6GK5	-60	1	60
Input impedance	grid	grid	cathode
Z in	high	high	low
Z in	Rg	Rg	1/gm
Z in (ohms)	1M	1M	91
Output impedance	plate	cathode	plate
Z out	medium	low	high
Z out	Rp	1/gm	>> Rp
Z out (ohms)	25k	91	25k

Where gm is transconductance, Rp is plate resistance, Rload is load resistance, Rk is the cathode resistor, and Rg is the gate resistor. These are the three modes of tube operation.

2. TRIODE 6GK5 CHARACTERISTICS AT LOW PLATE VOLTAGE

plate (V)	plate (Rp)	gm (S)	1/gm (ohm)	gain (mu)	plate (mA)
10	2770	0.0062	161	17.2	3.5
20	4660	0.0073	137	34.0	5.2
30	5330	0.0090	111	48.0	6.5
40	5450	0.0110	91	60.0	8.5

Where plate (V) is the plate voltage, Rp is the plate resistance, gm is the transconductance in Siemens (S), 1/gm is in ohms, mu is the amplification factor (gain), and plate (mA) is the plate current. This is the 6GK5 triode's characteristics (calculated) for low plate voltages (10V to 40V).

3. VACUUM TUBE OVERVIEW AND EQUATIONS

A vacuum tube is a voltage controlled device. Common grid has very high output impedance (parallel to load). Common plate is called a cathode follower. Common cathode has the highest power gain. A vacuum tube has very high input impedance. Tubes can be run at much lower plate voltage than specified. A vacuum tube has anemic voltage gain. Voltage gain is often made up for by using a transformer (10:1 or smaller) as the plate load. Current gain is very high. Input impedance relates more to the gate resistor (in parallel) to ground. The plate is also called the anode. Pentodes have higher values of plate resistance (Rp). Pentodes have a suppressor grid that repels electrons back toward the plate. Pentodes have a screen grid that reduces capacitances. This helps for high frequency (radio) amplification. Pentodes can be connected as, and used like, triodes. Voltage gain might be 15 in a triode; and 45 in a pentode. A normal pentode is a "sharp cutoff". Another type, called a "remote cutoff", has a control grid with uneven spacing. At a large bias, transconductance is lower. This limits gain so the tube won't be over-driven. Gain (transconductance) can be changed by changing the grid bias. As grid volts go more negative, plate current drops, transconductance drops, and the plate resistance increases.

```
Voltage gain = (mu x Rload) / (Rp + Rload)
     Rload is the resistor attached to the plate.
mu = gm x Rp    // mu is amplification factor
                // gm is transconductance
                // Rp is plate resistance
Output impedance is the parallel of Rp, Rload, etc.
Use as high a plate load as feasible.
Use a load of 10 times plate resistance to get near the "mu" value of gain.
A load of about 1 times plate resistance will get about 50% of "mu" in gain.
If driving another load make sure its impedance is within a factor of 10.
```

```
Fixed bias: negative voltage fed to the grid via a transformer or 1M resistor.
Self-bias: large grid resistor to ground.
        The cathode resistor creates a negative (relative) gate voltage.
Self-bias: also called cathode biasing (measure the voltage at the cathode).
```

```
Grid-leak bias: via grid leak or series grid leak (also detects AM signals).
Gird-leak bias: common values are 100 pF and 2M ohm.
Grid-leak bias: capacitance values range from 10 to 250 pF.
Grid-leak bias: resistance values range from 1 to 10 M ohm.
Grid-leak bias: is directly related to the amplitude of the input signal.
```

```
The 6418 and 6612 pentodes are used in common cathode mode.
     The screen grid controls output.
     The screen grid has compete control of the plate current.
     The screen grid can be used to control regeneration.
```

```
Child-Langmuir law: current is related to voltage to the power of 3/2.
     At low plate voltage, plate current is heavily reduced.
     The frame-grid 6GK5 can put out significant current at low voltage.
          For example: 3.5 mA of plate current at only 10 Volts of plate.
```

```
Rnoise = 2.5/Gm    // high Gm tubes (6GK5) are low noise tubes
```

1.13 Quick Biasing Appendix
THE RADIO BUILDING HOBBY
©2015

1. BJT: MPSA18 NPN

The emitter resistor is for stability. Voltage gain is the collector resistor divided by the emitter resistor. The bypass capacitor (sets the emitter to ground for AC) improves the voltage gain lost by the emitter resistor. For maximum collector output swing, the collector resistor should be 4.5 times the emitter resistor. To omit the emitter resistor and bypass capacitor, attach a large (see chart below) resistor from the wiper of a potentiometer to the base. Attach the ends of the pot to the plus power supply and ground. A capacitor is also needed to keep this voltage from leaking to ground via the tank's inductor coil. This capacitor can be located, and voltage introduced, below the tank. This capacitor may also be needed to remove DC from a previous amplifier section. The goal of the potentiometer is to be able to bias the transistor from saturation to cutoff. This can be used to increase positive feedback in a regenerative radio: since a control must exist somewhere. The base resistor (chart below) should be as high as possible as not to load the tank. This resistor is seen in parallel to the input resistance of the tank, which is very high. The values shaded below represent base resistor values that would negatively effect the tank's Q.

Rbase mA	1.5 Volts	9.0 Volts
0.25	1,300,000	14,800,000
0.50	650,000	7,400,000
0.75	433,333	4,933,333
1.00	325,000	3,700,000
2.00	162,500	1,850,000
5.00	65,000	740,000
10.00	32,500	370,000
20.00	16,250	185,000
50.00	6,500	74,000

2. JFET: J310 N-CHANNEL

Allow the gate to see ground. One method is via a tank's inductor. Another is by feeding the AC (radio or audio) signal through a capacitor to the gate and then attaching the gate to ground via a large 1 to 10M ohm resistor. Since the gate current is zero; the voltage at the gate is ground. A larger gate resistor will increase input impedance; which is in parallel to the AC signal.

The easy way to bias a JFET is self-biasing using a source resistor. The voltage at the top of the source resistor times -1 is Vgs. A typical J310 Idss is high: 43 mA. To bias near cutoff a high (10k-ohm) resistor could be used. A variable 10k ohm source resistor can be used as a set-and-forget type control to accommodate variations in specific J310 devices. Self-bias is susceptible to variations. To allow a half volt swing in a regenerative radio needs a ~800 ohm source resistor.

Attach the gate to ground, drain to +9V, and source through a 1M ohm resistor to ground. The source voltage is ~Vgsoff. For maximum current swing (often not needed), use a source resistor that will cause a voltage drop (measure it) of about 30% of the Vgsoff. A pot can be used and then replaced with a normal resistor. This means the current will be about 49% of Idss (chart).

3. VACUUM TUBE: TRIODE OR PENTODE

To bias using two components: use grid-leak biasing with the standard 100 pF capacitor and 1M ohm resistor. To bias using one component: attach the cathode to a 1N4148 diode to ground. The diode has a 1 Volt forward voltage drop; meaning the grid voltage will be set to minus 1V.

4. BIASING AND DEVICE OVERVIEW

On the left: BJT biasing uses either a voltage divider (ladder) or current (one resistor) at the base of the transistor. The emitter resistor is for thermal (**re**) stability. And the bypass capacitor increases voltage gain. It ties AC signals (radio and audio) to ground. Voltage gain is directly related to the load resistance. Current gain is ~500 (MPSA18). This is a current controlled device. A BJT offers positives of: very high transconductance and low power supply voltage (QRP). Its biggest disadvantages are: low input impedance and the possible thermal instability of the voltage gain.

In the middle: JFET biasing uses the source resistor to set a voltage that makes the gate, relative to the source, negative. A large gate resistor (1M to 10M) or tank circuit (DC is grounded via the inductor) is used to set the gate voltage to ground. Recall, there is no current flow through the gate. Meaning: the voltage at the gate is the same as the voltage across any resistance attached to the gate. The bypass capacitor increases voltage gain. The load size is directly related to voltage gain. Current gain is "infinite" for a JFET, ex. J310. This is a voltage-controlled device. A JFET offer positives of: very high input impedance and infinite current gain. Its disadvantages are: its highly variable Idds and Vgsoff parameters and its gate is more delicate than a base or grid.

On the right: triode or pentode biasing via a simple grid leak setup. This can also be used as a detector. The cathode can simply be attached to ground. Note that a tube can be biased in the same way as a JFET but a JFET cannot be biased using grid-leak. This is because the grid of a tube has some flow and can build up charge. The load size is related to the voltage gain. Current gain is very high for a vacuum tube. This is a voltage-controlled device. A vacuum tube offers positives of: high input impedance and very simple biasing. Its biggest disadvantages are: low voltage gain (remedied by using a 1:10 transformer as a load) and a heater that consumes power.

Current gain is fairly high on all these devices. However, on a typical BJT, with an hFE of 100, it is not as stellar. A purely current gain (no voltage gain) occurs when signal is applied to the base, gate, or grid but output via the emitter, source, or cathode. A purely voltage gain (no current gain) occurs when the signal is applied to the emitter, source, or cathode but output via the collector, drain, or plate. The base, gate, and grid are never outputs. And the collector, drain, or plate is never inputs. The emitter, source, and cathode can be both an input or an output.

Voltage gain on these devices is one when used as an emitter follower, source follower or cathode follower. The bypass capacitor increases voltage gain but at the cost of some noise. It is sometimes advantageous to have a stable but low voltage gain that is set by the load resistance divided by the resistance below the device. Meaning, using no AC (radio / audio) bypass capacitor.

And, as a special note, in radio, the input or output from these devices can be fed from the capacitances within the device. This is especially true when an emitter, source, or cathode RF choke pushes radio back through the device to the base, gate, or grid. Device capacitance is often the missing "component" in a non-apparent design. Another oddity is that a large load, especially in the collector, drain, or plate can cause RF (radio) to radiate and reappear at the tank. This is how a radio can actually be regenerative without resistive, capacitative, or inductive coupling.

1.14 Secrets of 1AD Radios
THE RADIO BUILDING HOBBY
©2015

1. TANK INPUT

Use a long wire antenna to gather RF energy. And introduce the radio energy to the tank using a tiny trimmer capacitance. Impedance matching is a complex subject. Two connected impedances should not be over 10 fold different or power will be lost. Incoming radio waves are at 377 ohms. On strong stations, loading the tank will make them louder. Using maximum capacitance (plates closed), make sure that the tank will tune the lowest frequency of the range: ex. 520 kHz on AM. If a tank does not tune far enough downward, add more turns to the coil winding. If a tank does not tune far enough upward, remove turns from the coil. Listen on a spotter radio (ex. DE1103) to know exactly where the radio is tuned. Markings can be made on the dial.

2. RF AMPLIFICATION

An ideal mode of operation to input radio from tank to amplification device is: common collector, common drain, or common plate. The very high input impedance will not load the tank. And the low output impedance can drive real loads: the tickler or positive feedback. The voltage gain of one will make for smoother regeneration. The number of tickler windings will be lower than the number of tank windings. The ratio will give a voltage gain, outside of the active device.

3. AF AMPLIFICATION

An ideal mode of operation for audio amplification after detection is: common emitter, common source, or common cathode. Primarily because power gain will be highest in these modes. Limit the audio gain by using a smaller bypass capacitor. Too much audio gain will cause problems in, and shut down, the RF section of the active device. Audio clipping should be avoided.

4. CONSTRUCTION TIPS

The physical construction should be rigid for all parts where radio (RF) is flowing. Keep all RF runs as short as possible. Designs should also keep RF away from the user interface: including all knobs and the headphones. Use a free Harbor Freight multimeter to measure the DC characteristics. A simple diode and crystal ear phone to ground can test any section with RF.

5. SPOTTER RADIO

A cheap "spotter" radio (DE1103) can be used on MW, SW, or FM. Tune the digital spotter radio, on the band of interest, to a specific station. Then over-regenerate the built radio and sweep the frequency until **silence** is heard (a carrier with no sidebands), instead of background noise. The regenerative radio is now tuned to that exact same station as the spotter. This can also be used for DX work. Find the station first on the spotter, over-regenerate, tune until the spotter hears silence. Then turn the spotter off and reduce regeneration. The station should be able to be heard.

6. EXPERIMENT

Do not be afraid to experiment. At 9 Volts a 1N34A only needs a 100 ohms in series. At 9 Volts the MPSA18 and J310 only need 500 ohms in series. Tubes like the 6GK5, 6612, and 6418 can easily handle 18 or 27 Volts of plate. Quarter watt resistors over 330 ohms are hard to damage at 9 Volts. Capacitors are often rated higher than 27 volts. The wire and toroids cannot be damaged. An audio transformer is 1420 ohms, meaning it has enough DC resistance to prevent damage to the other components. If any component becomes hot or starts smoking, disconnect the power. One problem that can happen is when a potentiometer is used to limit current. When turned to the lowest possible values the pot can start to smoke as too much current will be going through it. :)

7. MISCELLANEOUS

Use batteries for safety and no 60-Hz (AC) hum. Component values do not have to be exactly as shown in schematics: use what is available and close in value. Use mica (smaller value) and ceramic (larger value) capacitors. A gimmick capacitor is made of two twisted pieces of wire that still have their insulation on them: creating 1 pF to 5 pF. Typically, 0.001 uF or smaller is used to pass radio frequencies. And 0.100 uF or larger is used to pass audio frequencies. Capacitors are often marked in thousands (k) of picofarads (pF). Capacitor numeric code example: 104k is 100,000 pF or 0.100 uF. And 103k is 10,000 pF or 0.010 uF. And 102k is 1,000 pF or 0.001 uF. Resistor color code value are shown below: the third number is the number of zeros to tack on to the number. Resistor color code example: yellow-violet-red equals 4 7 2 or 4700 or 4.7k-ohms. And red-black-orange equals 2 0 3 or 20000 or 20k-ohms. And orange-orange-yellow equals 3 3 4 or 330000 or 330k-ohms. All the designs use linear potentiometers. On triode vacuum tubes, a common way to control regeneration is to vary the plate voltage. On pentode vacuum tubes, a common way to control regeneration is using the screen grid. The screen grid has complete control of the DC plate current. In the FM heliosdyne desgin a JFET is controlled in a similar fashion: by varying its drain voltage. The best size control potentiometer is 10k-ohms. These methods avoid using a throttle capacitor (expensive). The key to regeneration is to go smoothly in and out of oscillation. Feel free to increase or decrease the tickler winding to make this possible. On Armstrong (tickler) style regeneration, if no background hiss is heard, first swap the coil wires: there may be degeneration. Lower frequencies require less regeneration: it is harder to regenerate daytime shortwave. On a regenerative radio, to hear continuous wave (CW, Morse code), turn up regeneration until the circuit oscillates. Grid-leak resistors are, typically, from 1M to 5M. Grid-leak works by the capacitor charging on positive grid values and retaining a constant negative voltage. On tubes, the measured cathode voltage is also the grid bias (negative one times it). Using tubes, a change in grid voltage causes a change in plate current. Using a typical low-performance JFET (ex. MPF102), a standard value for a source resistor is about 1k-ohms with a 0.010 uF bypass capacitor. Choke values are often from 1 to 5 mH. A typical 8-ohm to 2000-ohm transformer will work but the voltage gain will be ~22 times lower than with a Bogen T-725 transformer. Instead of gathering costly parts, purchase an old, vacuum tube, **signal generator**. This will contain a metal housing, potentiometers with knobs, vacuum tube sockets, a dial, a vernier drive, and high-quality tuning capacitors. And other, costly, parts. A signal generator already has holes drilled for all the components. Metal shop is fun but a signal generator allows focusing on electronics. A vernier drive makes using a radio much more fun. The wavelength size in meters equals 300 divided by the frequency in Hertz. The frequency in kilohertz equals a million divided by two times pi (3.14159265359) times the square root of multiplying these together: the inductance, L, in microHenry and the capacitance, C, in picoFarads. Below are two charts. The left shows the resistor color code values. The right shows symbols and units for common powers of 10.

color	codes		unit	symbol	10 to the power of
black	0		giga	G	9
brown	1		mega	M	6
red	2		kilo	k	3
orange	3				
yellow	4				
green	5				
blue	6		milli	m	-3
violet	7		micro	u	-6
gray	8		nano	n	-9
white	9		pico	p	-12

1.15 The Radio Lab
THE RADIO BUILDING HOBBY
©2015

1. ANTENNA, TANK, GROUND

The first step to hearing shortwave signals on a home built radio is putting up an antenna. A safe way is to tape 25 to 50 feet of wire wrap (30 AWG) to the ceiling. Keep it away from anything electric. A tank is then built using a **T106-2** toroid (red) with 15 turns (an inner winding counts as a turn; cut 29 inches of wire) as the inductor (L). And a 360 pF variable capacitor (C) for tuning. Shoot for high quality because this tuning capacitor is the most important part of the radio. Get one with a vernier (reduction) drive and a dial. Tape can be used and station hot-spots marked. Ground can be, literally, a spike driven into the ground. However, for most simple radios it is often just a two inch piece of bus wire where the negative of the battery and all other "ground" lines are connected. It is important to make any lines with RF (radio) on them as short as possible. It is also best to isolate radio from the user interface (headphones, tuning dial, and regenerative control). This is done using RF chokes (RFC) and good circuit design. A mediumwave or local AM radio station tank will consist of a 60 turns on an **FT114-61** ferrite toroid (cut 74 inches of wire).

2. DIODE DETECTOR, CRYSTAL EARPHONE

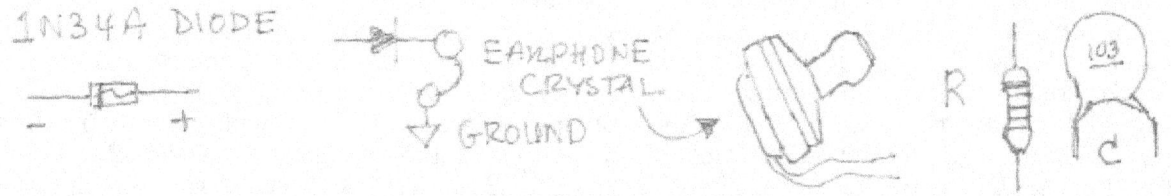

RF (radio) from the antenna will enter the tank using a small (10 to 500 pF) capacitor. This prevents loading the high impedance tank (with low impedance radio waves). The bottom of the parallel LC tank is attached to ground. The top of the tank will go directly into a 1N34A diode (+, anode side, triangle) and then the other end (-, cathode side, line mark) into a crystal earphone that is attached to ground. The variable capacitor is turned and a station may be heard. The tank above is for nighttime shortwave. To tune daytime shortwave less turns are needed. Less inductance and/or capacitance leads to higher tuned frequencies. The tuning capacitor can simply be soldered to another point on the toroid for daytime usage. Above are depictions of: the diode, an earphone, a resistor, and a capacitor. This is a crystal radio: listen in complete silence. A few of the most powerful stations (ex. Local AM or Radio Havana Cuba) may be heard faintly. Also try the other configuration shown below. **Always wear safety glasses or goggles when soldering.**

3. MEASURE VOLTAGE

Measure the DC voltage across (in parallel with) any component. Knowing the voltage, knowing the resistance, and using Ohm's law: the current across a component can be calculated. This current is the same for all parts in series. For example, if the collector resistor's current is known (Ic) then the emitter resistor's current is known (Ie): because they are (almost) equal. This is especially true with a high hFE transistor, like the MPSA18. Measuring voltage is safer and more convenient than opening a circuit to measure current (the meter must be placed in series with the component). If, by accident, the voltage is read across a 9 Volt battery, it may read "OL" or "overload". And, when the range is increased, it will read that 9 Volts. If, however, by accident, the current is read across a 9 Volt battery, there is no resistance. The current will be high and may permanently damage the digital multimeter. Pick up a free Harbor Freight multimeter. And measure the voltage drop across any component or measure the voltage with respect to ground.

4. MPSA18 NPN BJT TRANSISTOR

The graphic on the left shows why the capacitor (earphone) is attached to ground. At AC (RF or audio) there is no positive or negative as with DC. An AC source attached to ground enters the circuit and the output is simply returned to ground. The output of the collector is shown in the second graphic. The collector resistor ("R") is as if it is attached to ground and parallel to the 33k ohms the audio sees the crystal earphone as being. If the collector resistor is low, most of the energy (in parallel) enters the smallest resistance. This will take energy from the earphones. If the capacitor ("C") going to the crystal earphone were only 100 pF (5.3M ohms at 300 Hz) then nothing will be heard in the earphone: in series the largest resistance absorbs the most energy.

Above is the setup for a common-emitter amplifier. In step 1 the emitter is shown going to ground. In step 2 the collector is attached to a load and then to the positive power supply (the round circle). The voltage is seen across the load or at the point labeled "OUT". In step 3 a transformer ("T") is used with ~1500 ohms of DC resistance. The transformer allows attaching 8-ohm or 16-ohm earphones and boosting their impedance to 45k or 90k ohms. Often the "load" is determined by the components on hand. The load should be high for better voltage gain. In step 4, a large capacitor ("C") carries audio to a crystal earphone. Because the crystal earphone acts like a 16 nF capacitor it cannot be the load ("R") because it will not carry DC to the positive supply. At 300 Hz the crystal earphone has an Xc of 33k-ohms, dropping to 3.3k-ohms at 3000 Hz (audio).

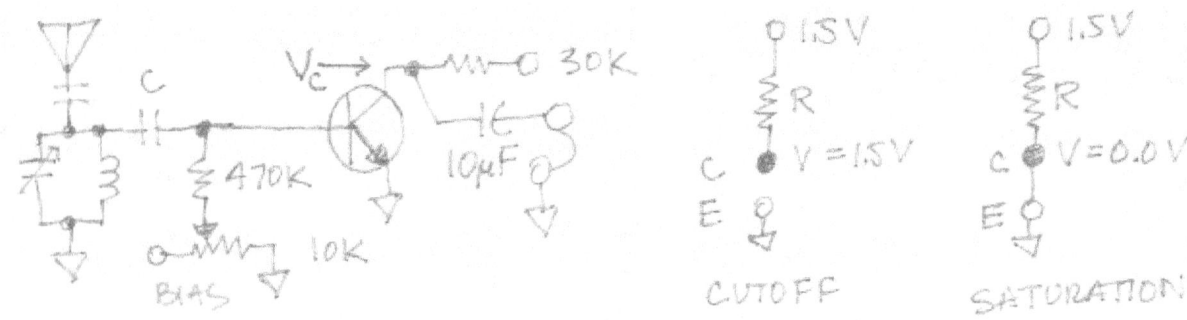

Above shows a hypothetical amplifier with a tank feeding AC (radio) into a capacitor ("C"). The capacitor keeps the bias current from reaching ground. The 470k bias resistor is attached to the wiper of a potentiometer whose ends are attached to ground and the +1.5V supply. As the bias increases the collector current increases and the internal emitter resistance decreases, thus increasing voltage gain. Attach a multimeter with the black lead to ground and the red lead to the collector to measure the collector voltage ("Vc"). If the leads are reversed the voltage will simply read "negative" volts. See the right graphics. If the collector voltage is close to the supply voltage (1.5 Volts), the transistor is in cutoff. It is an open switch. The voltage at the end of the resistor is the same as the supply. The leads of a multimeter measure voltage in this way with 1M to 10M ohm resistors that isolate it from the circuit. If the collector voltage is almost zero volts (0.2 Volts), the transistor is in saturation. It is a closed switch. This is how and why the voltage is inverted in a common-collector amplifier. With no applied base current, maximum voltage is output. And with maximum applied base current, minimum voltage is output. Ideally the capacitor ("C") will be about 1000 pF (.001 uF) or less: to load the tank. A small cap can be used to feed the antenna into the tank itself. Volume will decrease on strong stations but more stations can be tuned.

The output configurations shown above, the transformer or crystal earphones, are used throughout the following design builds. The Bogen T-725 transformer and Koss Sparkplug earbuds are ideal. However, two crystal earphones in series are inexpensive, sensitive, and immersive. The phone output is the same for a JFET, triode, or pentode: the difference is their biasing (see below).

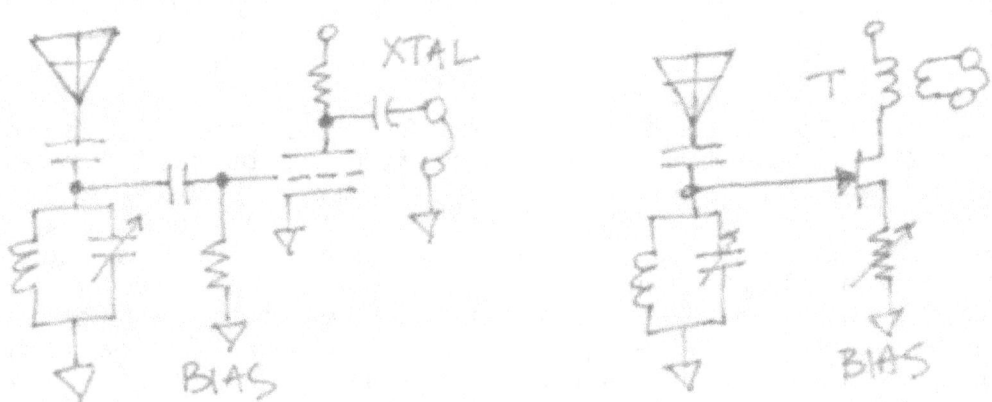

On the left is a triode amplifier, in common cathode configuration, with grid-leak bias. It uses a capacitor and a crystal earphone as output. Two crystal earphones can be used in series (better) or parallel. On the right is a JFET amplifier, in common source configuration, with source-resistor bias (self-bias). It uses a transformer and earphones. Output examples could be swapped.

Following this page are descriptions of the workings of: two BJT (MPSA18), two JFET (J310), and two vacuum tube (CK6612 pentode and 6GK5 triode) radios. Build these proven designs. The actual details about their build is in Chapters 3 and 4: the exact section will be noted. The final section of this chapter are advanced designs: the J310 JFET Heliosdyne FM radio and the 6GK5 triode Angelodyne SW radio. These take a little more skill to build and get working properly.

5. BUILD THE GLOBE PATROL JUNIOR: MPSA18 MW RADIO

See chapter 2, section 8 for complete build information.

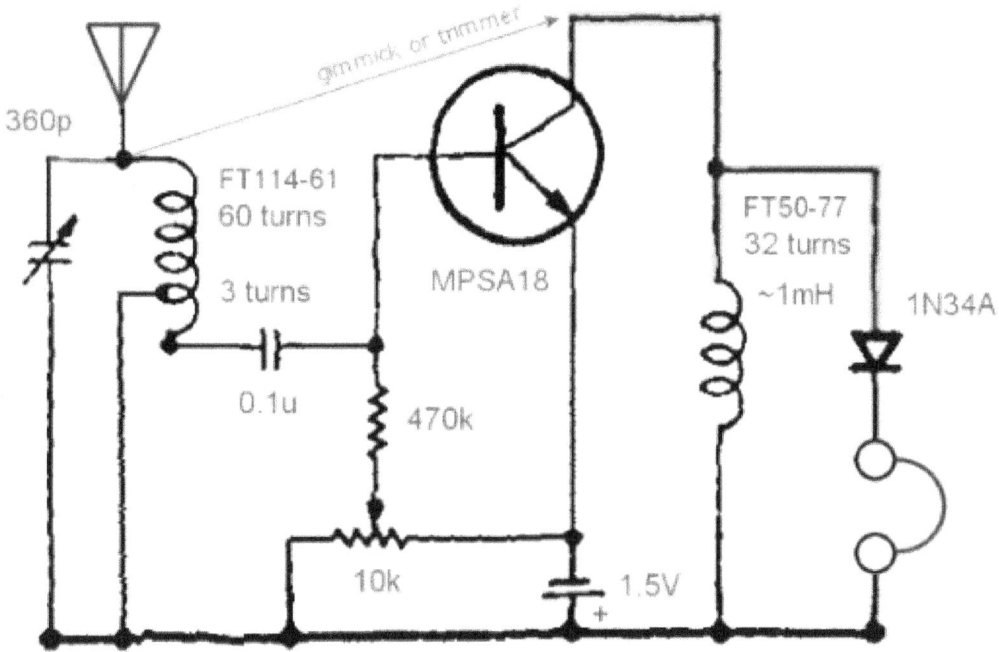

Above is the Globe Patrol Junior MW radio. Notice that the tank is grounded 3 turns before the end that is attached to the transistor's base. This inverts the input. RF travels through the 0.1 uF capacitor to the base. A 10k-ohm pot is attached to +1.5V and ground. The pot's wiper goes through a 470k-ohm resistor to reduce current. The 0.1 uF cap prevents DC from touching ground via the tank inductor. Increasing the base current will increase transistor gain. This is common emitter mode. The only collector load, at DC, is a few ohms (wire) in the RF choke. The antenna gathers RF and the tank selects the tuned frequencies. Energy is inverted 180 degrees by the turns below ground and enters the base. The turns ratio of 60 to 3 provides a 20 fold increase in current. This is, indirectly, "tapping down" on the coil. The highest voltage is at the top of the coil. However, a BJT transistor is a current-controlled device. RF is output at the collector but it cannot pass the 1 mH RF choke. Half the RF enters the diode, gets detected, and is heard via the crystal earphone. The other half of the RF will radiate and is picked up by the tank. This normally does not work because collector output is 180 degrees out of phase. A gimmick, or two twisted wires, with low (1 to 3 pF) capacitance can be used to directly attach the collector to the tank. This is a regenerative radio. The heart of the design was taken from the Science Fair Globe Patrol radio. The original extra bands and two audio amplifiers were omitted. The original emitter resistor and its bypass capacitor were, also, omitted. With a 1.5V supply, a 470k-ohm base resistor, and a gain of 500, current will only be ~1 mA. The transistor will dissipate some of that 1.5 mW. Without the pot, the transistor needs an emitter resistor for better stability. The MPSA18 is used as an RF amplifier. If the transistor is set to detect audio, it is lost to ground through the 1 mH coil. Below are diagrams showing the RF (radio), AF (audio), and DC workings of the Globe Patrol Junior.

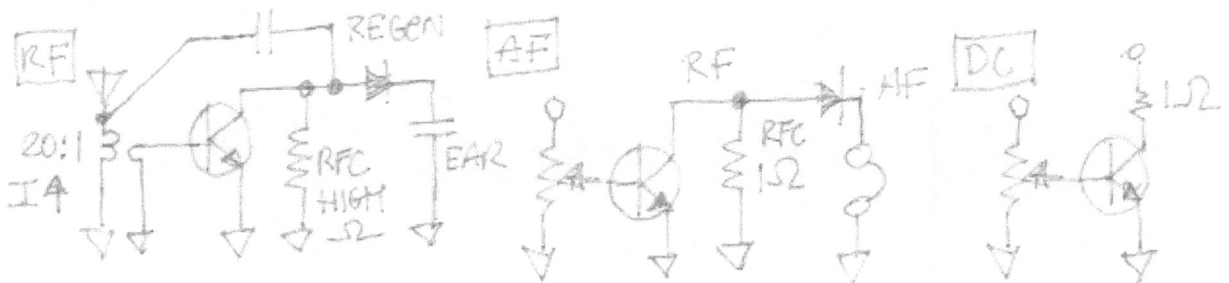

6. BUILD THE DEE/MITCH-DYNE II: MPSA18 SW RADIO
See chapter 2, section 3 for complete build information.

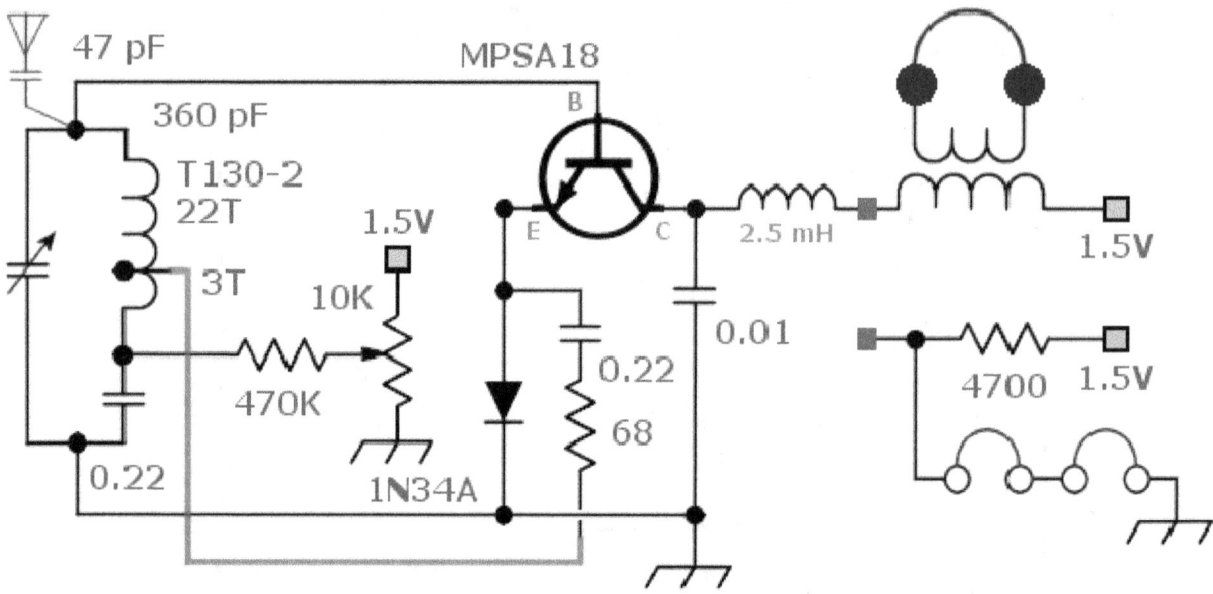

Above is the Dee/Mitch-Dyne II SW radio. The antenna gathers radio waves and passes them to the tank via a 47 pF capacitor. Notice the 0.22 uF capacitor below the tank inductor. This is large and has no effect on the tank: no resistance to radio waves. The RF that is selected by the tank enters the transistor's base. At the collector, RF is kept from the earphones via an RF choke. Radio is shunted through a 0.01 uF capacitor to ground. At RF, this is common collector. RF exits the emitter and half is dumped to ground. Half of the RF, trapped above the diode, passes through a large 0.22 uF capacitor and a 68-ohm resistor and directly into the tank (3 turn tap). The resistor increases input impedance. A common collector's output is in phase with the input and can be directly added to the tank (Hartley style). This is regeneration. Regeneration is controlled at the base. A 10k-ohm pot is attached to 1.5 Volts and ground. The pot's wiper goes through a 470k ohm resistor: providing under 1 mA of current. The tank's tuning capacitor and the additional capacitor below the tank's inductor prevent DC from touching ground. It, instead, goes to the base. The pot controls the base current. At DC, the emitter sees the diode as resistance. DC cannot enter the tank due to the 0.22 uF capacitor. At the collector DC passes through the RF choke, through the ~1400 ohms of wire in the T725 transformer, and then to +1.5V supply. At audio, the emitter sees a small diode resistance. Audio can, also, pass through the 0.22 uF capacitor, 68-ohm resistor, the tank inductor, and the 0.22 uF capacitor, to ground. The collector sees a high resistance at audio: 16k ohm headphones are boosted to 90k ohms via the transformer. This gives good audio voltage gain. This is common emitter at audio. When crystal earphones are used a 4700 ohm resistor replaces the transformer and audio is shunted directly to ground via the 16-nF capacitance of the crystal earphones. Below are diagrams showing the RF (radio frequency), AF (audio frequency), and DC (direct current) workings of the Dee/Mitch-Dyne II SW radio.

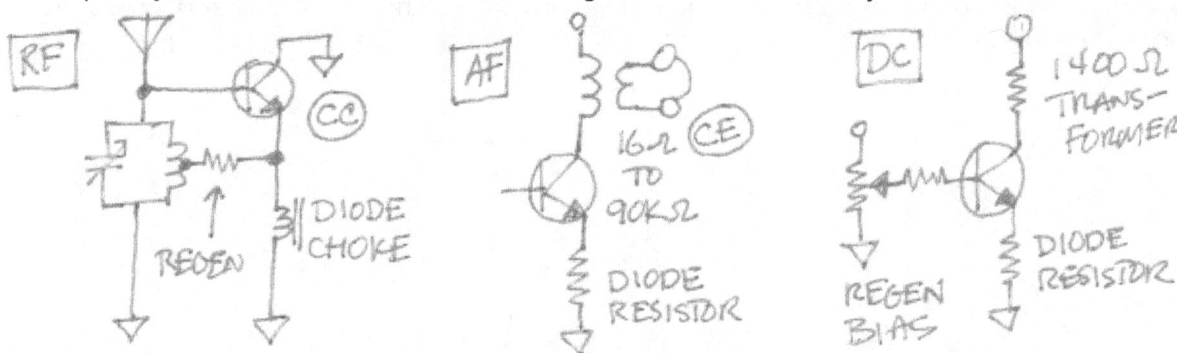

7. J310 N-CHANNEL JFET TRANSISTOR

The input impedance of a JFET is very high and will not load a tank. The current gain is "infinite" in that little current flows. The Achilles heal of the JFET is mediocre voltage gain. Below is the result of testing nine J310 JFET devices. The first column is the Idds or maximum current that occurs with a gate-to-source voltage of 0 Volts. The second column is Vp or the pinchoff voltage: current is zero. At the bottom the average was calculated: 44mA for Idds and -6.12 Volts for Vp. The standard deviation (SD) is also given. In a normal distributions, the rule of thumb is that 68% of values will fall within one SD; 95% will fall within two SD's; and 99.7% will fall within three SD's.

JFET	Idds	Vp
1	39.8	-8.40
2	45.7	-6.58
3	42.9	-5.22
4	42.8	-4.96
5	45.9	-7.09
6	44.3	-6.87
7	44.3	-5.53
8	44.2	-5.56
9	46.3	-4.88
average	44.0	-6.12
SD	2.01	1.19

Vgs		Vgs		Id	Id
0.0	Vp	0.00	1.00	Idss	44
0.3	Vp	-1.84	0.50	Idss	22
0.5	Vp	-3.06	0.25	Idss	11
1.0	Vp	-6.12	0.00	Idss	0

The average device will have the above right values, for plotting in a transfer curve.

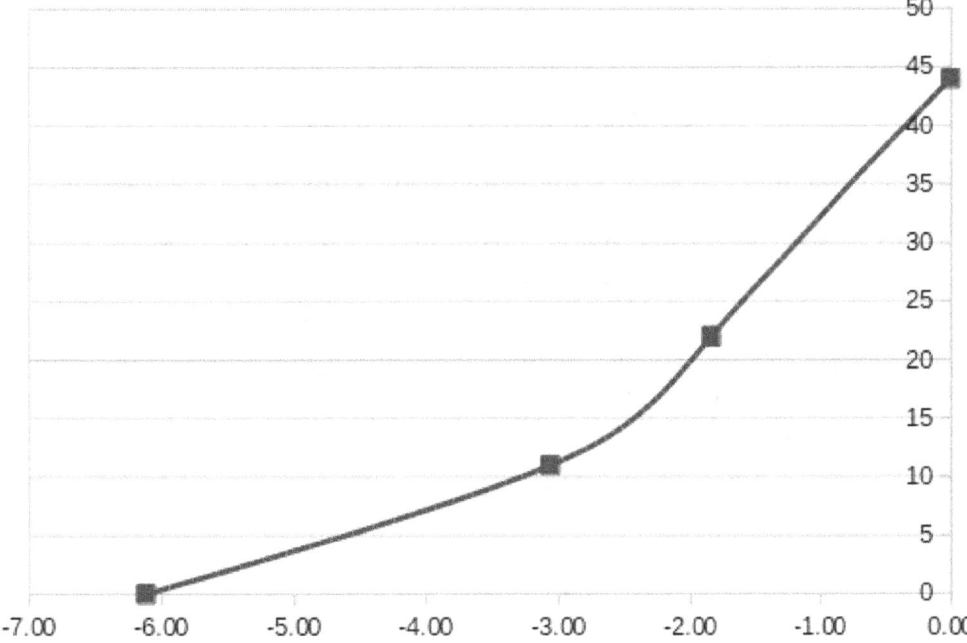

Above on the x-axis is the Vgs or voltage gate-to-source in Volts (negative numbers). On the y-axis is Id or the drain current. If a half volt of swing is needed, the values for the resistor will be 2.5 mA at about -5.5 Volts. Using Ohm's law, this comes out to a value of 2,200 ohms. The 2.2k ohm source resistor sets the bias for the circuit. A 10k-ohm potentiometer can be used to set the value; because each device differs. The voltage drop across the drain resistor can be measured and the current calculated via Ohm's law. The simplest voltage supply for a JFET is a 9 Volt battery. And this is still not ideal, as many of those volts will end up across the source resistor for biasing.

8. BUILD THE SCIENCE FAIR RADIO: J310 MW/SW DETECTOR

See chapter 2, section 7 for complete build information.

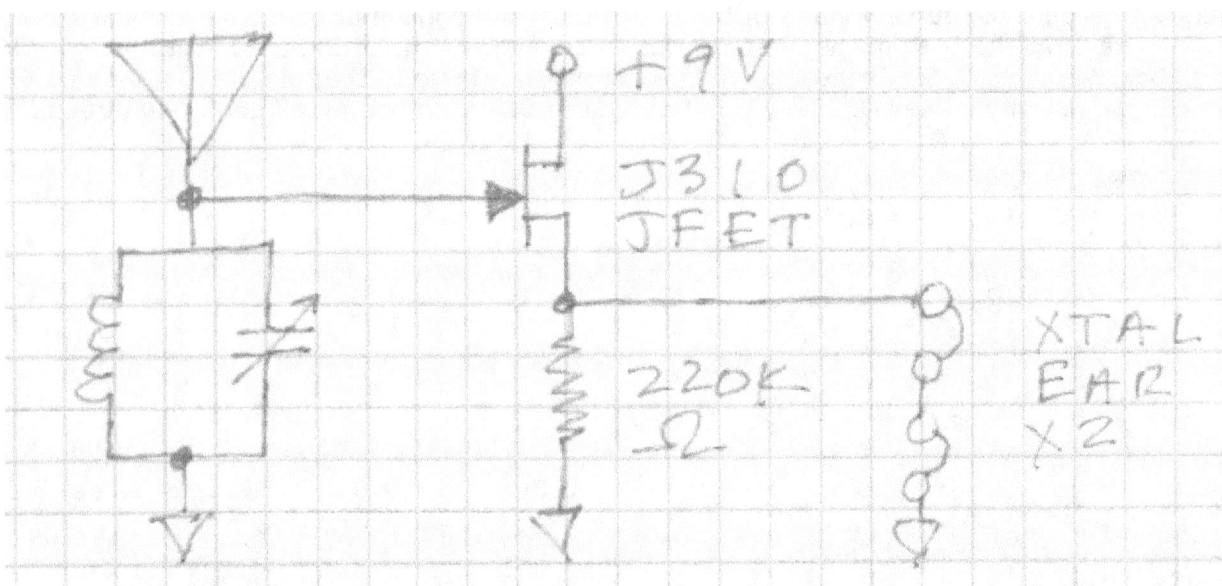

Above is the Science Fair MW or SW detector. The antenna gathers radio (RF) and sends it directly to the tank. The tank selects the frequencies of interest and passes them to the J310's gate. All other frequencies, fall through the tank, to ground. The JFET is setup in common drain or source-follower mode. The drain is attached directly to the +9 Volt supply battery. Input impedance is extremely high and will not load the tank. However, the direct antenna connection does cause some loading. With this simple setup, strong stations will be heard at good volume. At DC, the gate is attached to ground via the tank coil. The 220k-ohm source resistor sets the gate-to-source voltage to be at (near) cutoff. The voltage above the resistor is the value of cutoff. DC cannot flow through the crystal earphones: as they act similar to a 16 nF capacitor. At the tuned radio frequencies, it is as if the tank is missing. At the source of the JFET, the tuned RF sees ground through the crystal earphones. At RF the phones are 20 ohms total: 10-ohms each. At audio, a single earphone acts as a 33k-ohm load: 66k with two in series. One or two crystal earphones can be used. Their resistance is lower than the 220k ohm biasing resistor, so the audio mostly appears across the phones. The detector works because the JFET is biased to cutoff and current will only flow during positive peaks. The conversion gain has no voltage gain but has "infinite" current gain. This is significant and helps drive the real-world crystal earphones. Note that it is usually a good idea to use an electrolytic (100 uF) audio capacitor before the crystal earphones, so as not to expose them to direct current (DC). Below are diagrams showing the RF (radio frequency), AF (audio frequency), and DC (direct current) workings of the Science Fair MW/SW detector.

The Science Fair MW/SW detector is not a regenerative radio. However, with as little as seven parts (one earphone), it is possible to hear local MW and strong shortwave stations. This is a great radio for science projects or to get people interested in radio. The radio bug can bite hard.

9. BUILD THE JFET HELLENEDYNE: J310 SW RADIO

See chapter 2, section 6 for complete build information.

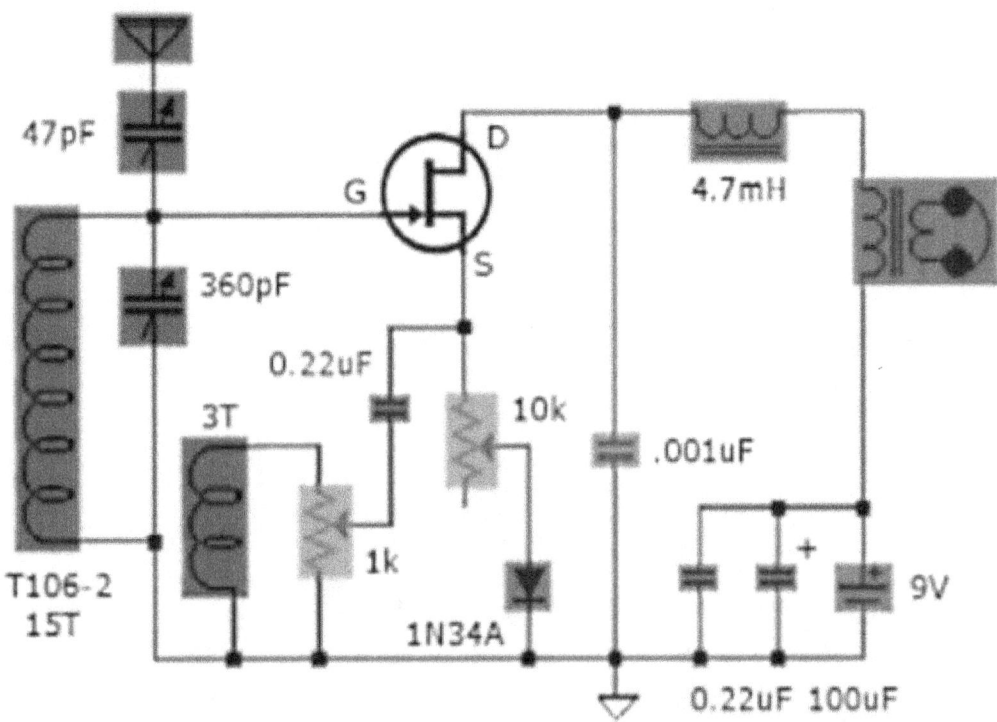

Above is the JFET Hellenedyne shortwave radio. The antenna feeds the LC (inductive-capacitative) tank through a small, variable, trimmer capacitor. This helps to not load the tank. A small amount of RF energy is often ideal. The tuned radio frequencies enter the J310 JFET's gate. At the drain, RF cannot pass the 4.7 mH RF choke and is shuttled to ground via a 0.001 uF (1000 pF) capacitor. At RF the circuit is setup as a common drain: also known as a source follower. RF is output from the transistor at the source (S). The 10k-ohm potentiometer sets the DC operating point for the JFET. Specifically, this source resistor sets the bias point. The 1N34A diode sets the gate at at least 0.3V or 300 mV. The diode also acts as an RF choke. Even at low 10k-ohm potentiometer values, at least half of the RF energy will enter the 0.22 uF capacitor on the source branch. The 0.22 uF capacitor leads to a 1k-ohm pot's wiper. The ends of the pot are connected to ground or through a tickler coil (3T or 3 turns; the tank is 15 turns on a T106-2 shortwave toroid). The 1k-pot dumps RF to ground or through the tickler. This is Armstrong style of regeneration.

Audio, at the transistor's source can see ground through both arms. If the 10k-ohm pot is low enough in value, audio can pass through the diode. And, if not, the tickler coil is of no resistance to audio. And, at low regeneration, the pot dumps audio directly to ground. Low source resistance will increase audio voltage gain. Audio energy from the transistor's drain is blocked from ground because the capacitor is too small: only 0.001uF. Audio readily passes through the 4.7 mH RF choke. And then enters the audio transformer: 90k-ohms at 300 Hz. Since voltage gain is related to the drain load (90k) divided by the source load (~1k), decent voltage gain is achieved. Without the transformer, the 16 ohm Koss Sparkplug earphones would hear nothing. At audio the circuit is setup as common source for maximum power gain. The output is via the drain (D).

There are two capacitors in parallel with the battery. The electrolytic capacitor (note its positive pole in the schematic) ensures that current is available. The smaller capacitor allows stray RF energy to pass to ground. At DC, the gate is grounded through the tank's coil. The 10k-ohm pot sets the gate's voltage: the DC bias point can be changed to accommodate varying JFET parameters. The gate-to-source voltage can be varied. The drain, at DC, sees the audio transformer as ~1400 ohms of resistance; because it is, essentially, a very long piece of wire.

10. VACCUM TUBES: CK6612 PENTODE AND 6GK5 TRIODE

11. BUILD THE OSCILLODYNE: CK6612 SW SUPER-REGEN

See chapter 3, section 6 for complete build information.

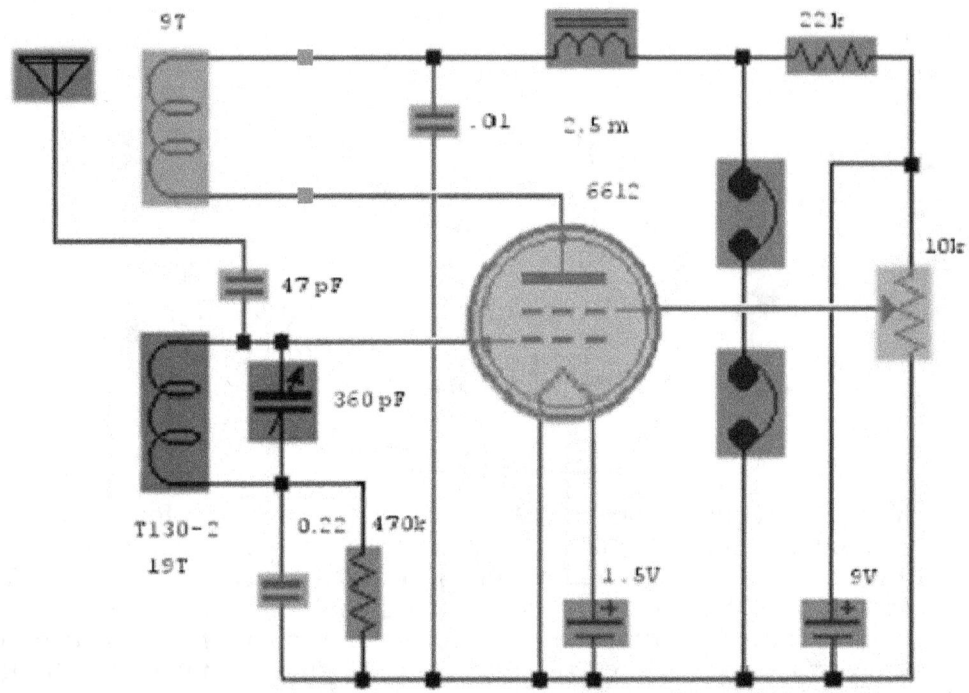

The original Oscillodyne was designed by J. A. Worcester, Jr. in the 1930's. This is a modern version using a CK6612 miniature (pencil), pentode vacuum-tube. This type of tube is used in common cathode mode: it does not have a separate cathode. Radio energy is fed to the tank via a 47 pF capacitor. A 1.5 Volt battery powers the filament: where electrons boil off. The plate runs a tickler coil; after which, energy is shunted to ground via a 0.01 uF capacitor. This is Armstrong style regeneration. The 2.5 mH RF choke keeps RF energy out of the two, series, crystal earphones. A single earphone can be used as well. DC from the plate sees the tickler and choke as almost zero ohms. DC enters a 22k ohm resistor that is hooked to +9 Volts of power. Super-regeneration is controlled via a 10k potentiometer to the screen grid. It is a better design to put a 10k-ohm resistor on the wiper (to limit grid current) with an RF decoupling capacitor. The tube's bias is provided by a grid-leak resistor and capacitor: located below the tank. It could have been placed above the tank as well. DC cannot enter the earphones because they act like a capacitor.

This circuit looks exactly like a regenerative radio. What makes it a super-regenerative radio are two minor changes. First, is the small value of the grid-leak's resistor. The resistor is usually 1M to 10M ohms; but here, it is only 470k-ohms. Second, the tickler coil is excessive: 9 turns: which is about half the size of the tank's inductor. These two changes cause a disruption of the grid circuit, in a rhythmic fashion. The excessive tickler tends to cause electrons to build up on the grid. This shuts the circuit down; but not for long as the 470k-ohm resistor drains electrons off the grid. The circuit then fires up again. Gain is about 1 million or 120 dB. Selectivity is not as good as a regenerative radio. But the loudness of the big stations is outstanding for a small, tube circuit.

12. BUILD THE HELLENEDYNE: 6GK5 SW RADIO

See chapter 3, section 2 for complete build information.

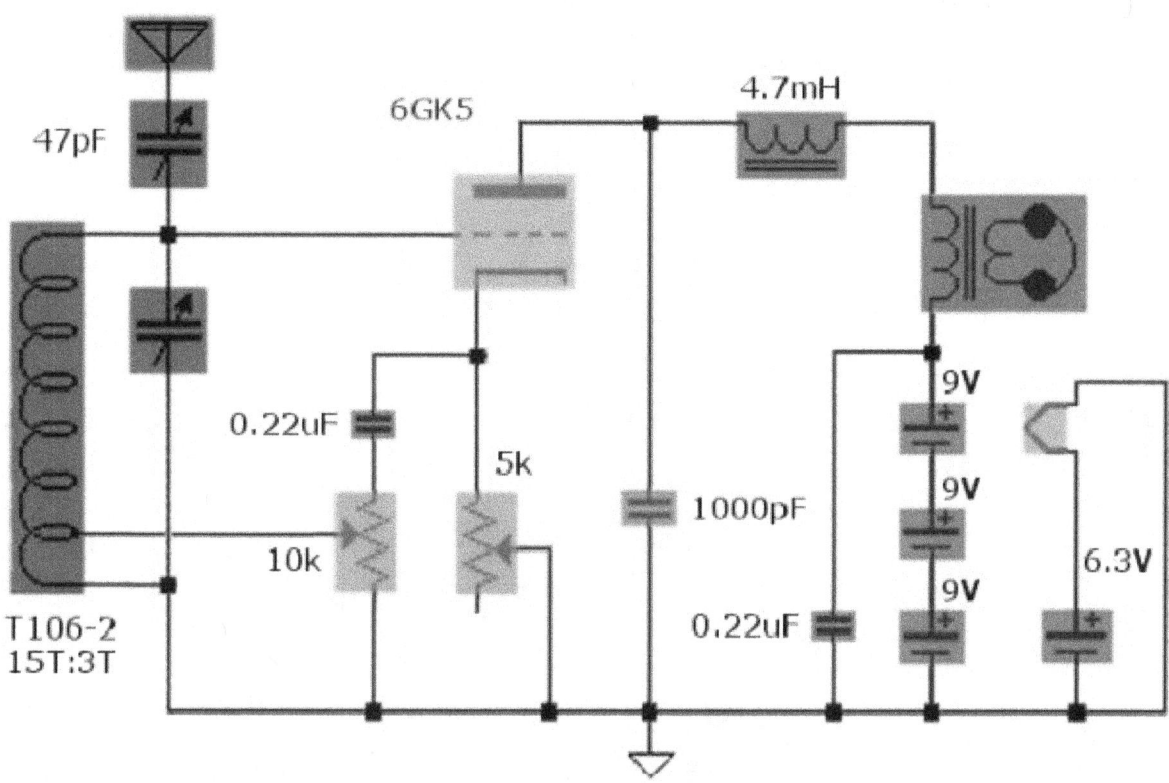

Above is the Hellenedyne shortwave radio. The circuit is very similar to the JFET Hellenedyne above. Radio energy enters a variable 47 pF capcaitor and then enters the tank and grid. The tank dumps all frequencies to ground, except those being tuned. The grid sees ground through the tank's inductor. The 5k-ohm cathode potentiometer controls the bias by placing the cathode at a certain voltage (variable) above the grid. At the plate, radio energy is shunted to ground through the 1000 pF capacitor; blocked by a 4.7 mH RF choke. The circuit is common plate or a cathode follower at RF. The RF choke keeps radio out of the transformer and headphone circuit. The heater, that boils off electrons, is powered via a 6.3V source. A 6V lantern battery will suffice. The plate is run on three, series, 9V batteries: for 27 Volts total. The 0.22 uF capacitor at the top of the batteries shunts and stray RF to ground. RF energy exiting the cathode goes into a capacitor and a 10k potentiometer. The pot is also hooked to ground. The pot's wiper goes directly into the tank (3 turns from the bottom). This is Hartley style regeneration. Energy in a cathode follower is non-inverting and can be directly injected into the tank. At audio, the circuit is setup as common cathode for high voltage gain. The voltage gain is slightly decreased because of the position of the 10k-ohm and 5k-ohm potentiometers. The RF, AF, and DC paths are shown below.

1.16 Advanced Designs
THE RADIO BUILDING HOBBY
©2015

1. BUILD THE HELIOSDYNE: J310 FM RADIO

See chapter 2, section 1 for complete build information.

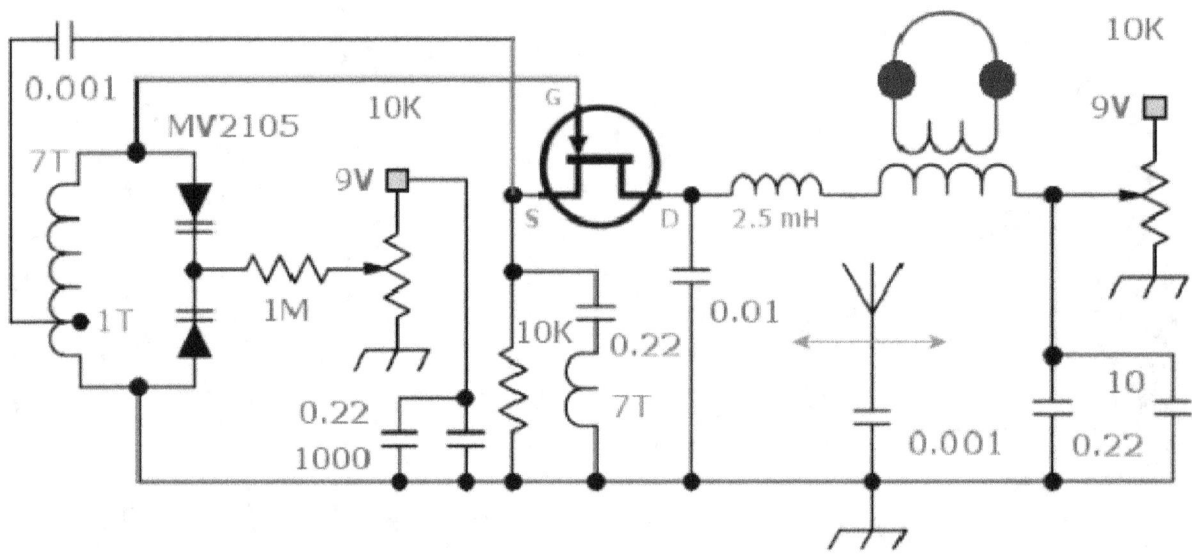

Above is the Heliosdyne FM radio. The tank uses two varactors instead of a small tuning capacitor. The varactors' capacitance is determined by the voltage applied to the 1M ohm resistor. This resistor, physically close to the varactors, isolates the tank from the tuning mechanism. The left 10k-ohm pot has 10 turns and allows for razor-sharp tuning. The only pitfall is that tuning is related to battery voltage. The two capacitors at the 9 Volt source shunt stray RF to ground. This radio is so sensitive that the antenna is just located in the vicinity of the tank. Moving it too close to the tank will shut the radio down. The frequencies selected by the tank enter the J310 JFET's gate. At the drain, RF is kept out of the phones via an RF choke. RF is shunted, via a 0.01 uF capacitor to ground. At RF the mode is common drain or source follower. RF emerges from the source. It cannot reach easily ground through the 10k-ohm resistor or 7-turn coil. RF is shunted through a small 0.001 uF capacitor to a 1-turn tank tap. This is Hartley style regeneration. Regeneration is controlled via varying the voltage using a 10k-ohm (1-turn) pot (right). Two capacitors shunt stray RF to ground and supply current during transient peaks. The RF choke isolates the regeneration control from the RF. At the source, audio sees a 10k ohm resistor. Voltage gain here is only ~9 (90k divided by 10k). However, audio may pass through the 0.22 uF capacitor (212 ohms at 3400 Hz; 2.4k ohms at 300 Hz) and easily through the 7-turns of wire (~0 ohms; FM RF choke). Voltage gain is now 90k divided by 212 ohms or ~424 (at 3400 Hz) or ~37 at 300 Hz.

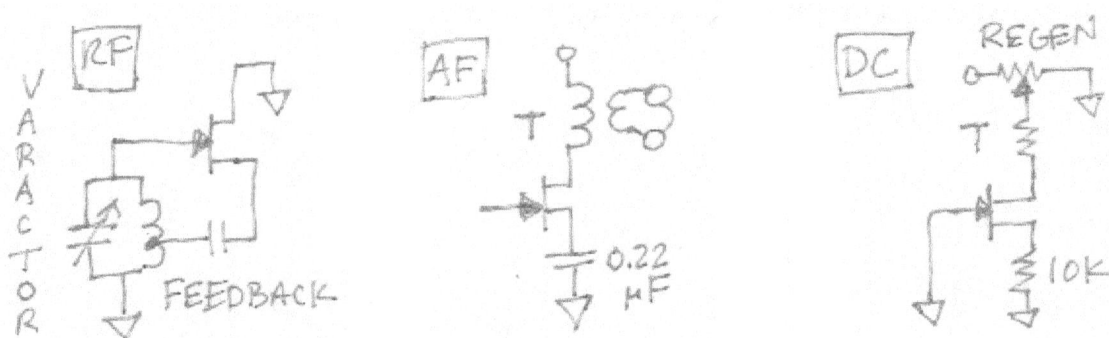

2. BUILD THE ANGELODYNE: 6GK5 SW RADIO

See chapter 3, section 1 for complete build information.

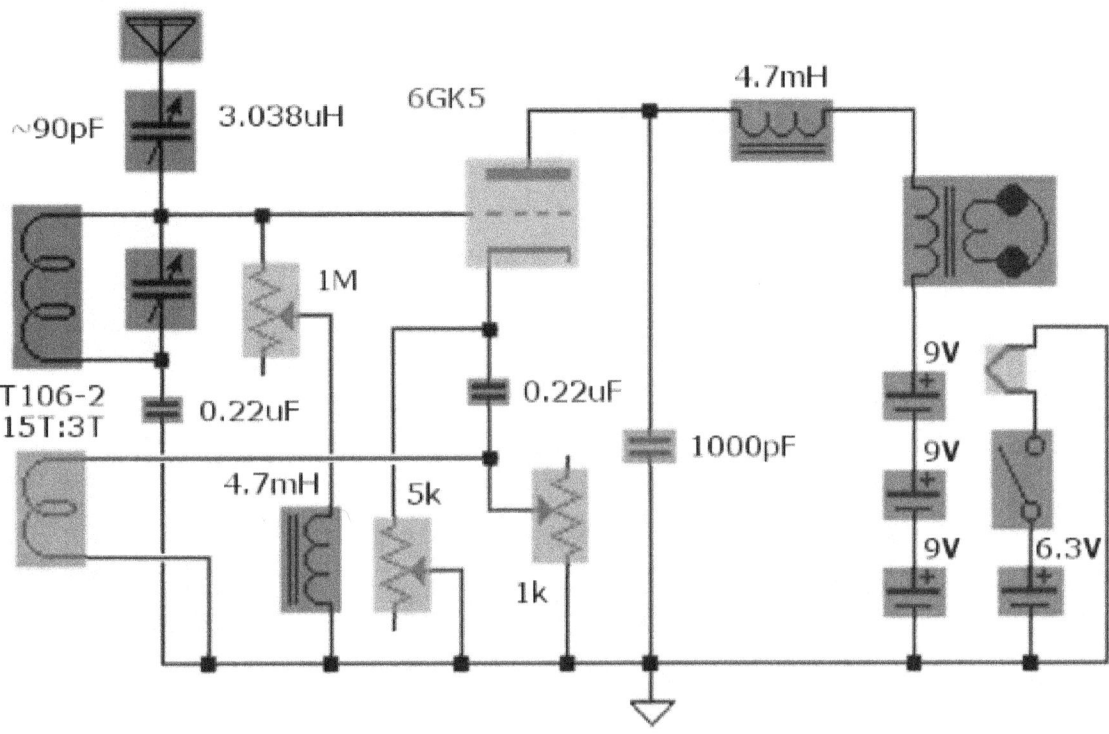

Above is the Angelodyne shortwave radio. This circuit is similar to the Hellenedyne above: only the differences will be noted. Regeneration style is Armstrong instead of Hartley. A 3-turn (3T) tickler coil is used instead of a 3-turn tap. The regeneration control is a 1k-ohm potentiometer, instead of a 10k-ohm. This design can be used as a regenerative radio or as a super-regenerative radio. The former is better for DX work; while, the later is good for listening to strong stations.

When the 1M-ohm potentiometer is near zero ohms, or connected directly to ground, then the radio is in a regenerative mode. The triode's bias is set by the 5k-ohm cathode potentiometer.

The 0.22 uF capacitor, below the tank, prevents DC from reaching ground through the tank's inductor coil. The 90 pF capacitor is set low to allow a tiny amount of radio energy to enter. The circuit is very sensitive in super-regeneration mode. Electrons or negative charge will build up on the grid and shut the circuit down. The 1M-ohm resistor is adjusted such that this negative voltage is bled off and restarts the circuit. The 5k-ohm cathode potentiometer sets up a bias. This mode only works in a vacuum tube due to electrons striking the grid on their way to the plate.

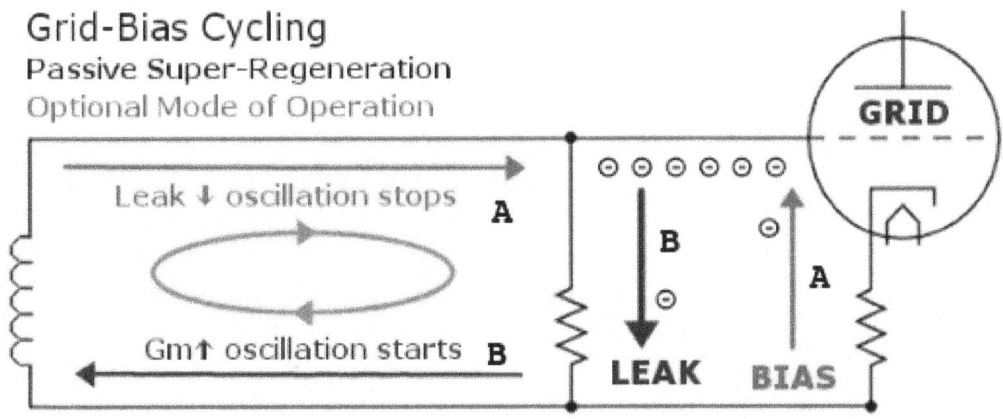

3. EXAMINE THE SHORTWAVE SPONTAFLEX

Below is a circuit called the Shortwave Spontaflex by Sir Douglas Hall. It appeared in the April/May 1993 issue of a magazine called RADIO BYGONES: page 18. Four tuning inductors (Lx4) and four regenerative capacitors (x4) have been omitted for clarity. How is this circuit working?

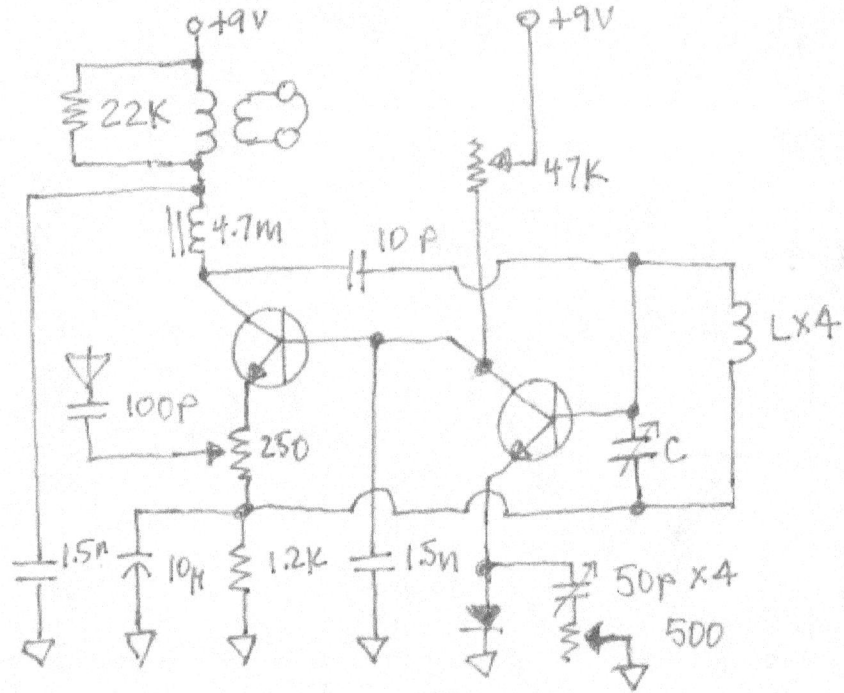

The circuit will be evaluated. The first step is to determine the direct-current (DC) flow. DC does does not flow through any size capacitor. All capacitors are removed: leaving the circuit shown below. The 500-ohm resistor, below the 50 pF variable capacitor, also disappears.

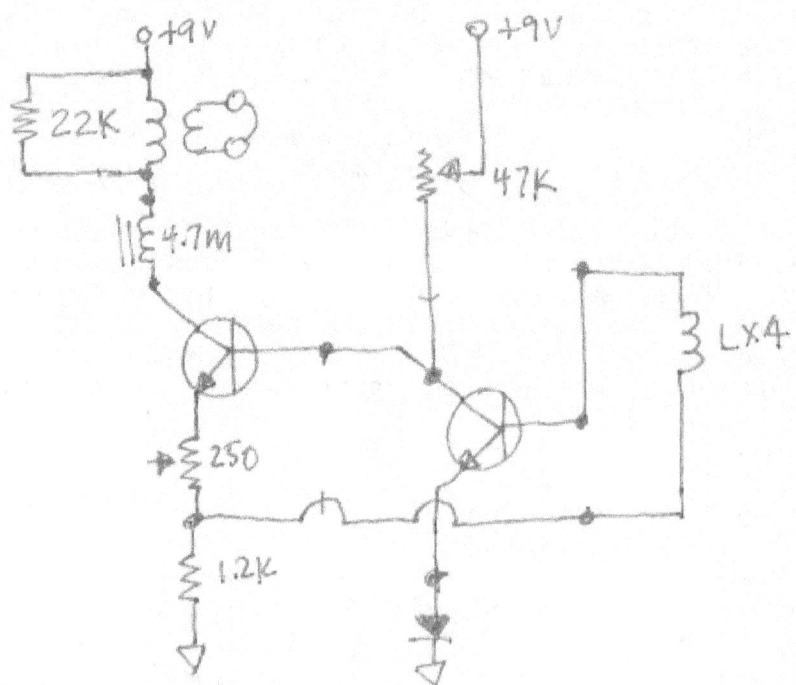

The next step is to examine the other components at DC. The transformer primary is ~2000 ohms. The 4.7mH RF choke and RF tank inductor (L) are about 0-ohms. The diode varies but is a small resistance. These changes are shown below. The DC paths are becoming clearer.

The graphic below summarizes the circuit at DC. The left transistor: a 9V battery is attached through ~2k-ohms to the collector. The emitter is attached to 1.45k-ohms to ground (250 plus 1200). The base current (and gain) is variable via a 47k-ohm potentiometer. The right transistor: a 47k-ohm pot attaches a 9V battery to the collector. The emitter sees ground through a diode (small R). The base is fed DC via the emitter ladder from the first transistor: for stability.

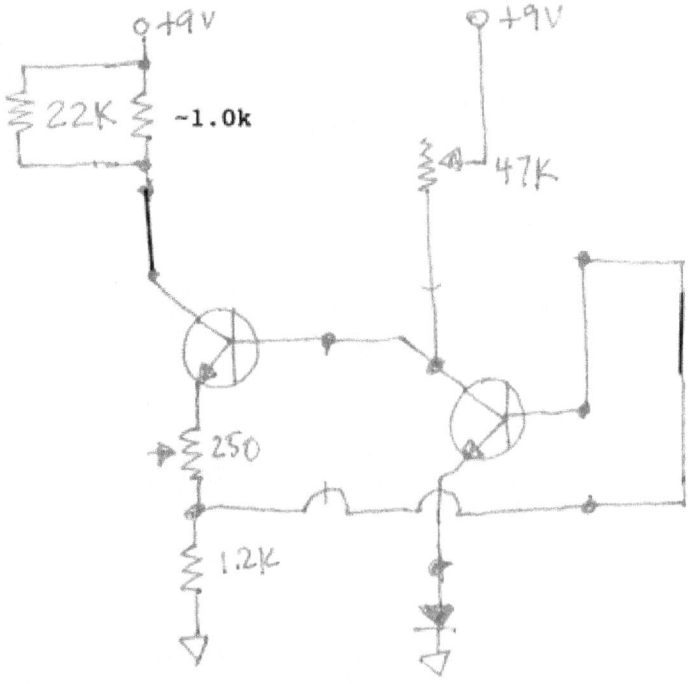

Next the radio (RF) path is visualized. RF cannot pass the audio transformer. RF will not enter the 1.2k-ohm resistor because the 10uF capacitor is a direct path to ground. RF will also not go through the 47k-ohm pot because that point is connected to ground via the right 1.5nF capacitor. RF will not pass the 4.7mH RF choke; but, if it does, the left 1.5nF capacitor will dump it to ground. The 100 pF antenna cap limits the RF to the emitter. The 250-ohm pot, turned toward the 10uF cap, dumps all radio energy to ground. In other positions, energy enters the emitter. The base sees ground through the right 1.5 nF capacitor. This is a common-base amplifier. Voltage-amplified radio energy will exit the collector. It then travels through the 10 pF (very small) capacitor to the LC tank (tuning). The 10uF capacitor attaches the bottom of the tank to ground. Radio enters the base of the right transistor. The collector sees ground (right 1.5 nF). This is a common-collector amplifier. Current amplified RF exits the emitter. The 10pF variable cap (set once) and 500-ohm pot control regeneration. Audio is created at the emitter of the right transistor.

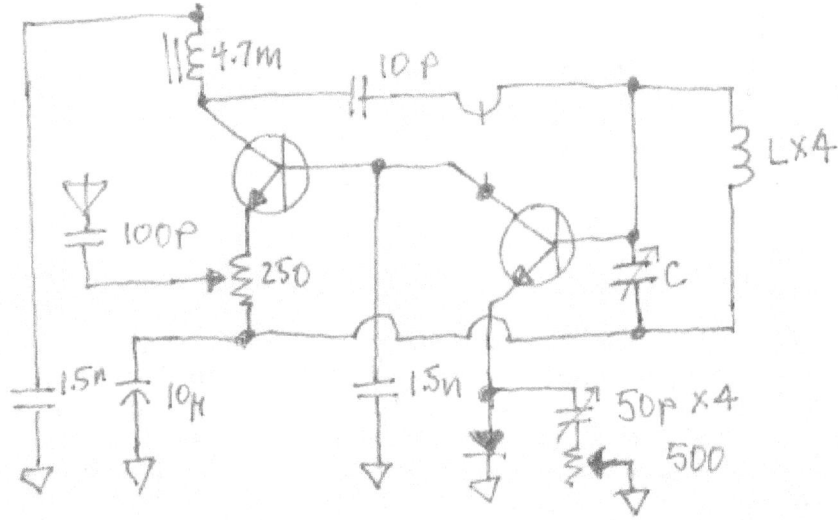

The next step is to determine the audio path. All lines with small capacitors (up to 1.5nF here) will be erased. The following graphic is what is left. The audio path is becoming clearer.

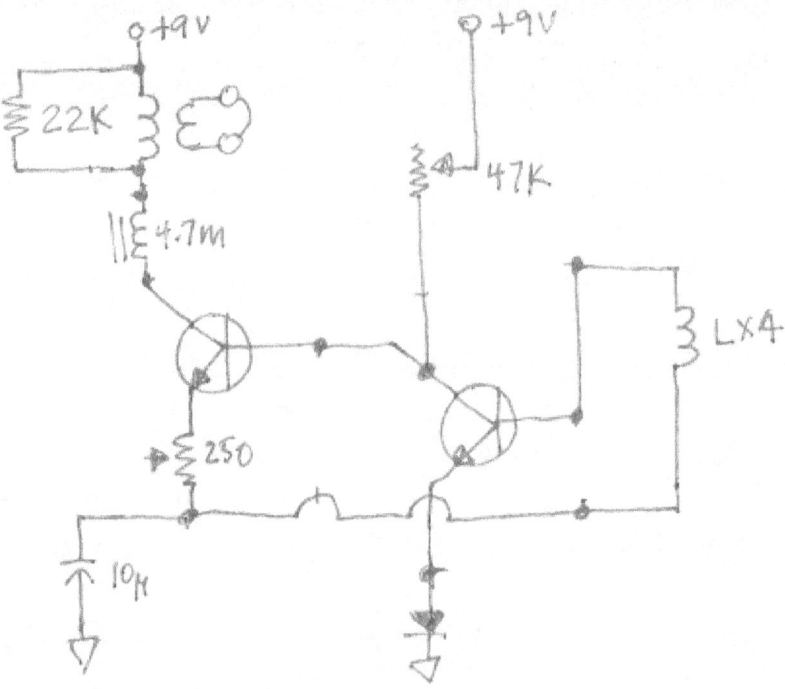

The tank inductor (Lx4), the 4.7 mH RF choke, and the 10uF capacitors are all low resistances at audio. The 22k-ohm resistor is for stability: RF cannot cross the transformer. The audio transformer can boost 8-ohms to ~2000 ohms. The graphic below shows what is left.

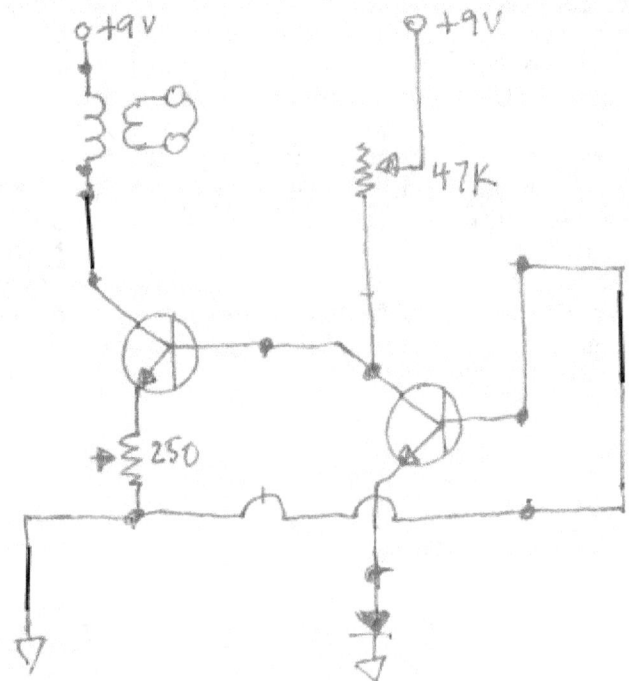

Audio is created at the emitter of the right transistor. The base sees ground (through the 10uF capacitor). This is a common-base amplifier. Voltage amplified audio emerges from the collector and passes to the base of the left transistor. This is a common-emitter amplifier using a 250-ohm (pot) emitter resistor for stable amplification (not dependent on re). The voltage and current amplified audio appear across the ~2000 ohm transformer and is transferred to the 8-ohm phones. By looking at each component (DC, radio, audio) separately, the circuit workings emerge.

1.17 Extra
THE RADIO BUILDING HOBBY
©2015

1. PIN DIODE SWITCHES

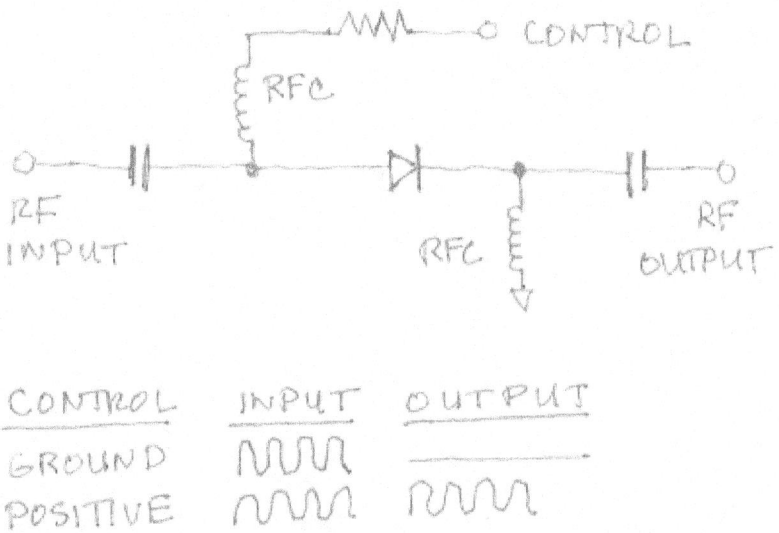

PIN diodes are diodes with an **I-region** (undoped intrinsic region) between the P-doped and N-doped regions. PIN diodes are used as an electrical switch with 40+ dB of isolation. When OFF PIN diodes offer high resistance to radio frequencies. When ON they offer low resistance to radio frequencies. PIN diodes are an alternative to real switches and select a path for radio to flow. For example, PIN diodes could be used after an antenna to select between multiple band-pass filters.

Above shows the usage of a PIN diode. At radio, the capacitors offer no resistance. The RF chokes (RFC) block radio. Therefore, at radio, the only obstacle is the PIN diode itself. The PIN diode might need 1.3 Volts of DC applied to turn on. When the control line is at ground level or unconnected, radio cannot pass the PIN diode. However, when the control is at a high enough positive voltage, DC flows through the resistor (current limiting), through the diode, and to ground. The capacitors block DC and the RF chokes offer little resistance to DC. Once the 1.3 Volts of DC is applied, radio frequencies can readily pass. And the RF at the input will be seen at the output. Note that the RF chokes could be replaced with high value resistors as they, too, can block radio.

2. AIR VARIABLE CAPACITORS

Below shows how to identify the three types of air-variable capacitors: (A) straight-line capacitance, (B) straight-line wavelength, and (C) straight-line frequency.

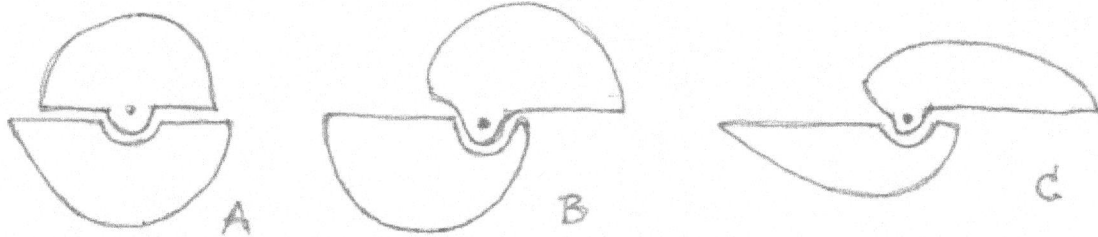

3. TESTING JFET GATES

Resistance testing (ohms) was able to determine if a JFET's gate was blown. The gate is supposed to be isolated (via a high resistance) from both the source and drain. Connect the black lead to the gate (left lead, while laying flat) and the red lead to the source (middle lead) or drain (right lead). The reading should be millions of ohms. If its low (< 2000 ohms), the JFET is blown.

Test	Expect	black-red	Blown 1	Blown 2	Blown 3	Blown 4
resistance	3M	GD	108	1890	117	611
resistance	3M	GS	108	19	121	579

G = gate, D = drain, S = source

2.1 Heliosdyne
Single Transistor FM DX Receiver
©2011

1. INTRODUCTION

There are two popular single-transistor FM radio designs on the net. The first, called Andy Mitz's One Transistor FM Radio, appears on his *FM-Only Radios* site. Andrew's design: is made of 24-parts, exploits common-gate JFET super-regeneration, tunes using two variable capacitors, and uses a TL431CLP as an audio amplifier. The TL431CLP is a programmable shunt regulator. It is an active device consisting of an on-chip voltage reference, operational amplifier, and NPN transistor.

The second FM design, called The Radio Shack Special, was designed by Sir Pedro. Pedro's design: is made of 21-parts, exploits common-gate JFET super-regeneration, tunes via one variable capacitor, uses a home built radio choke and gimmick capacitor, has two potentiometers, and uses an LM386 as an audio amplifier. The LM386 is a low voltage audio amplifier that contains about 10 transistors. For more information visit Patrick Cambre's site: *Set Sail with Sir Pedro and Patrick...*

2. HELIOSDYNE

The Heliosdyne is a solid state version of my Angelodyne detector. The Angelodyne can be a *super-regenerative drain detector* capable of FM demodulation. The Heliosdyne was designed as a single transistor FM radio that uses no extra active devices for audio amplification. It was made for hearing distant or weak signals, so-called FM DX. The radio uses 21-parts and consumes 2 mA, or 18 mW. The Heliosdyne, differing from Andy's and Pedro's radios, has no printed circuit board, no mechanical tuning capacitor, and no custom gimmick capacitor (as with *The Radio Shack Special*).

At radio frequencies the Heliosdyne uses common-drain (aka source follower) which makes for very high input impedance, about 100M ohms. The output is of low impedance and high current. This is ideal for adding energy back into the tank. The tank itself consists of a home built inductor: seven turns at a diameter of 10 mm and a length of 37 mm. The coil can be bent to fit the range. For the other seven-turn inductor use a pencil as a form and compress it as tightly as possible.

The tank uses two varactor tuning diodes (MV2105: rated 6.8 pF to 30 pF, Q = 400) being fed via a 10-turn potentiometer. Tuning is so precise that the stereo and high energy mono channel can be resolved separately (if the amplitudes are sufficient). The 1M ohm resistor insures that no RF energy will be flowing in the tuning control. There is no hand capacitance or skipped stations.

At audio frequencies the Heliosdyne uses common-source for maximal power gain. Stations are not loud but easily heard. Many stations can be listened to hour-after-hour. An RF choke keeps the Bogen T725 audio transformer and regen control clean of radio energy. The two 16-ohm Koss Sparkplug earbuds are paralleled (8-ohms) and brought to 40k ohms via the T725. A hobbyist with sound-powered phones will have ~18 dB on my setup. The antenna, 25' of wall-mounted wire wrap, is grounded and moved from 5 cm to 1 cm away from the tank's end. Re-tuning can be necessary.

Unlike MW or SW, FM is unforgiving of sloppy design. The receiver should be rigid and built exactly as shown. The placement of all but the controls, transformer, and headphones are critical. Keep everything, including ground, away from the tank. The J310 (J309 is preferred) JFET is static sensitive. So avoid touching the gate (or ground), solder the gate last, and check the circuit before adding power. If distortion is noticed, the JFET's gate may be damaged.

To use the Heliosdyne: turn the pot to the end with the high pitched squeal. Reduce power until the squeal disappears. Or use the point before the sound stops. Or another, below the silence. Tune to a weak station or background noise to fine adjust. Avoid loading the tank with the antenna.

The Heliosdyne can be miniaturized by using a small transformer and trimmer pots. The 0.22 µF capacitor and inductor in parallel with the source load form an *audio voltage gain enhancer*. The enhancer can be removed if the 0.001 µF source capacitor is increased to 0.22 µF. Performance is optimal as shown. The varactor system can be replaced with a trimmer cap. Use fresh batteries or regen will be harsh. Piezos can be used instead of earphones. See the Angelodyne Detector section.

3. SCHEMATIC

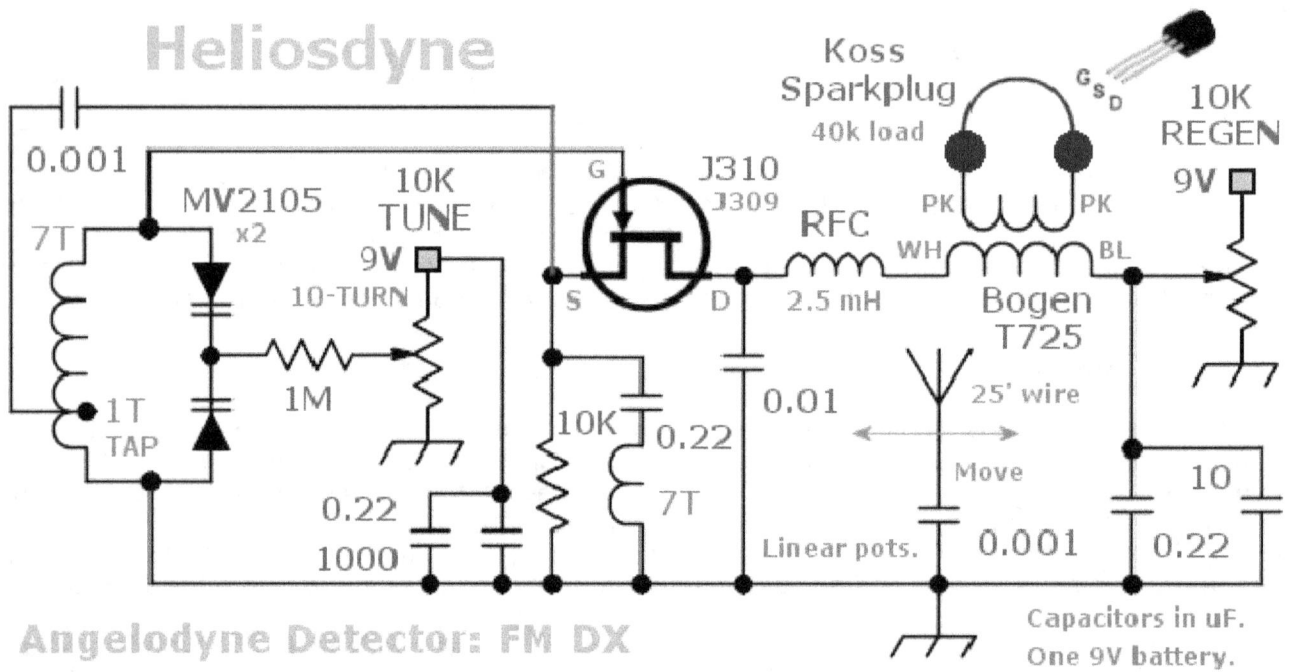

4. DISCUSSION

The Heliosdyne is a super-regenerative design that has enable me to pursue the hobby of 1AD FM DX. This is a prototype and subject to fine tuning. I released the Heliosdyne early due to its performance. In my first sitting, after the radio was running, I logged 42 FM stations. And this was without searching dusk or dawn, or varying antenna coupling, or using a directional dipole. I'd like to thank Andrew, Pedro, and Patrick for their fine work. **My hope is that the Heliosdyne design will inspire hobbyists to undertake a new hobby: 1AD FM DXing.**

The Heliosdyne uses slope detection. This is an AM detector, offset tuned to f1 or f2 (as seen above, right), used to detect an FM signal (f, as seen above, left). In general, an FM broadcast band signal can be slope detected at plus (f2) or minus (f1) 0.050 MHz (50 kHz) from the station frequency (f). The range over which slope detection is often possible is from about plus or minus 0.020 MHz to 0.080 MHz. A wide AM filter is desirable (ex. 40 kHz). Some stations can only be detected at the lower end of the range (0.020 MHz) due to the steep slope of the FM signal.

PHOTOS OF THE HELIOSDYNE FOLLOW

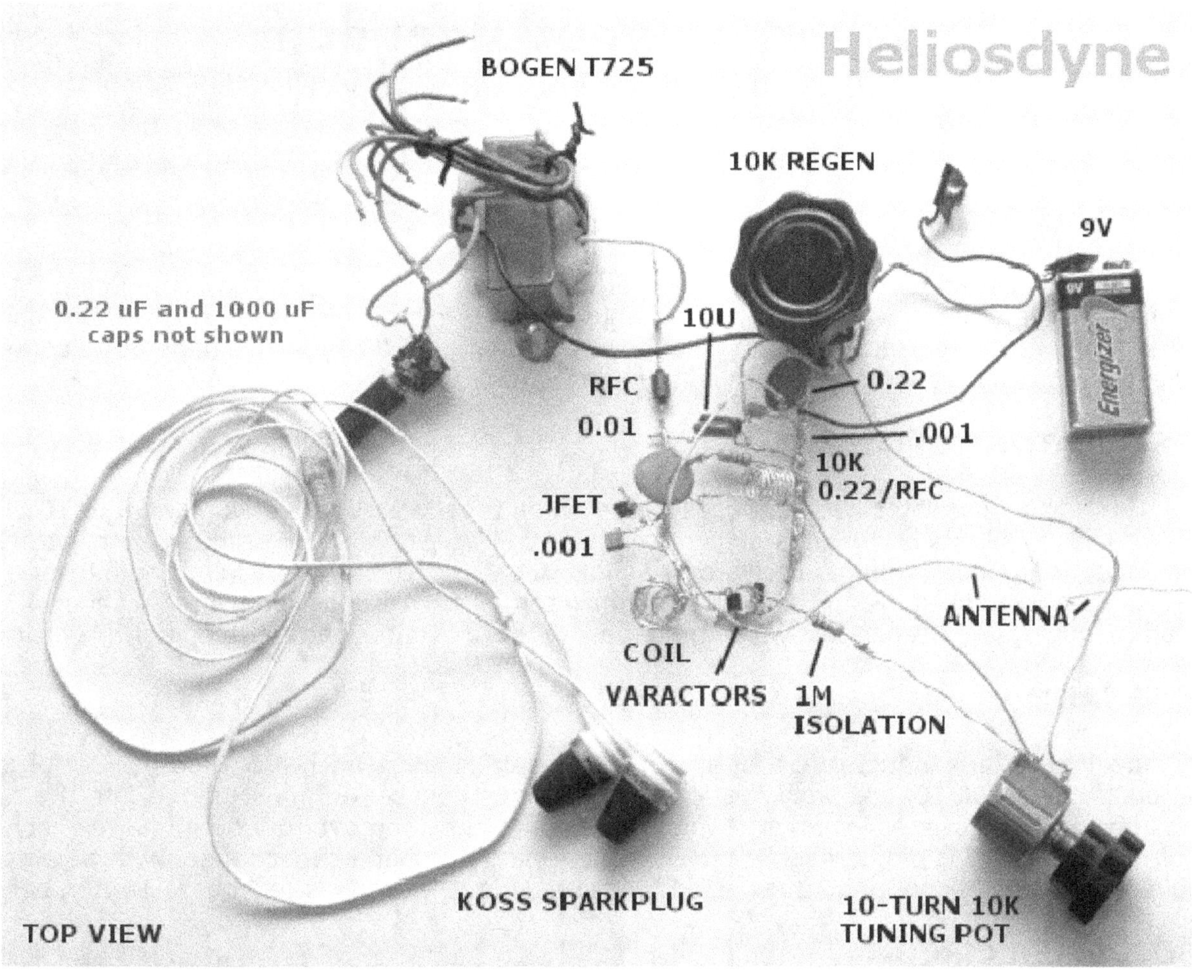

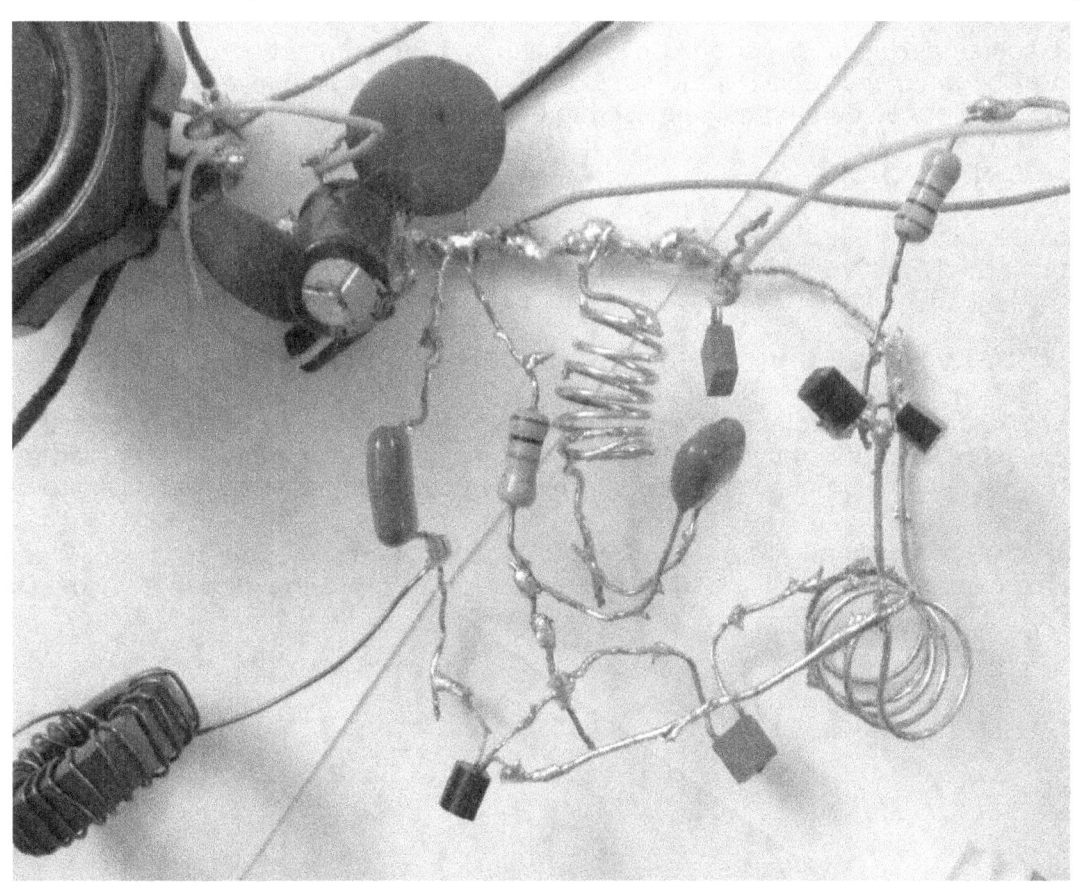

BOGEN T725

Heliosdyne

10K REGEN

9V

0.22 uF and 1000 uF
caps not shown

10U

RFC
0.01

0.22

.001

10K
0.22/RFC

JFET
.001

ANTENNA

COIL

VARACTORS

1M
ISOLATION

KOSS SPARKPLUG

TOP VIEW

10-TURN 10K
TUNING POT

2.2 Heliosdyne Addendum
Single Transistor FM DX Receiver
©2011

1. INTRODUCTION

This is an addendum to the Heliosdyne article that describes a single JFET radio for FM DX. This single active device (1AD) radio is adaptable for QRP (low-power) usage on the ham bands. The Heliosdyne is a departure from the ubiquitous common-gate JFET super-regenerative detector. The Heliosdyne is a regenerative detector setup as common-drain or *source-follower at RF*.

2. OSCILLATOR TOPOLOGY

The Hartley oscillator was historically used on all broadcast bands including 88 to 108 MHz FM. That said, a Colpitts oscillator is preferred above 50 MHz. With a 47 pF trim capacitor attached from source to ground: a Colpitts oscillator is formed. A capacitative divider is created between the 47 pF air-variable capacitor and the J310's Csg or internal *source-gate capacitance*, about 4.1 pF. The tank's inductor and capacitors will need to be altered (this is beyond the scope of this article).

3. THE TANK

Ideally the tank inductor has an air core. Toroidal FM coils wrapped on iron powder of μi=6 (example Amidon T50-10) are problematic in 1AD designs because RF must be fed into the tank. This can be achieved using a *gimmick* (two covered wires entwined over a length of a few inches). Use of real capacitors is problematic in terms of fine tuning and hand capacitance. With a varactor, combining a 1-turn 10k pot and 1-turn 1k pot, in series, may be used instead of the 10-turn pot.

4. FOLLOWER CIRCUIT

The source circuit must perform several actions, including setting up the DC load, feeding RF back into the tank, and reducing the audio load. I found two unique ways to accomplish these tasks. The principles can undoubtedly be used to boost performance of other hobbyists' receivers.

A. DIODE RF CHOKE

A diode will pass half the RF, DC, and audio. This means a diode acts as a perfect RF choke that traps half the RF above itself. A 1N34A diode has 200 ohms of resistance at a forward voltage of 0.50 Volts and 100 ohms at 0.80 Volts. This fact can even be used to control a Colpitts oscillator.

B. AUDIO VOLTAGE GAIN ENHANCER (AVGE)

An AVGE or audio voltage gain enhancer allows low resistance at audio but blocks both the DC and RF energy. The AVGE boosts voltage gains on many 1AD common-collector, common-drain, and common-plate audio amplifiers. This is due to voltage gain being inversely related to this load.

The AVGE is simply an AF capacitor (ex. 0.22 μF) in series with an RF choke. When the DC resistor is high enough, the RF coil itself is not that critical because the RF will "see" the DC load.

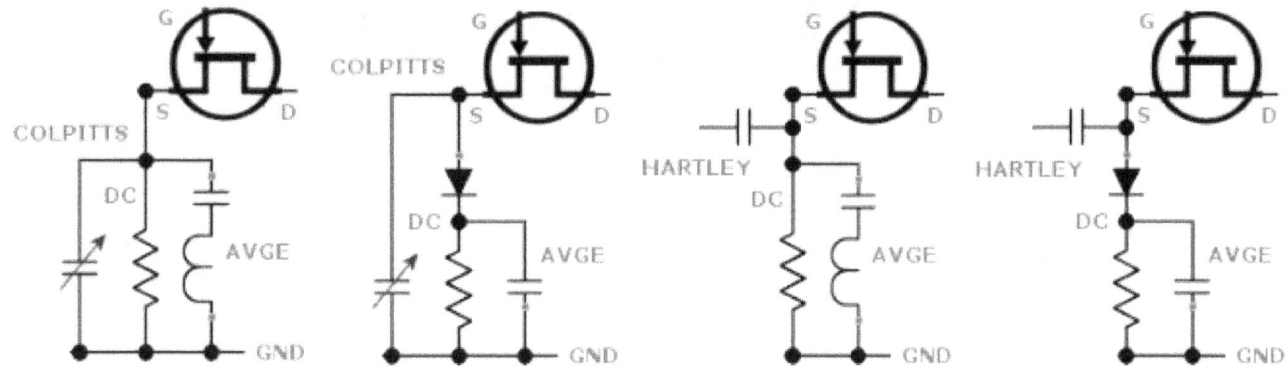

Also consider that a 0.22 µF capacitor in series with, say, 250 nH is resonant at ~680 kHz. This means it will not only boost audio but tend to dump medium wave energy out of the system. The AVGE will distort audio somewhat and likely can be omitted when using sound-powered phones.

5. QUENCH WAVEFORM

Add a 500 ohm potentiometer under the *audio voltage gain enhancer* to alter the quench waveform (as per *Charles Kitchin*). This was not fully tested but may be needed for narrowband FM.

6. VOLTAGE STABILITY

The Heliosdyne prototype had one serious design flaw. With usage, as the battery voltage dropped, the varactors could no longer tune the upper end of the band. One solution is the use of a *zener diode*. The power used to drive a super-regenerative detector alters *selectivity* (push less power) and *sensitivity* (push more power). RF chokes with high DC resistance can be problematic.

7. PHONE CIRCUIT

The use of sound powered phones will result in a 18 dB gain over the $16 Koss Sparkplug. The Bogen T725 transformer is advised as it provides 1420 ohms at DC and ~90k ohms at audio.

8. ACTIVE DEVICES

An n-JFET is electrically similar to an n-channel depletion-mode MOSFET. Higher gains may also be possible using JFETs with higher gfs or IDDS. Take into account the device capacitances.

9. MISCELLANEOUS

Ideally the AVGE inductor is at a 90 degree angle with the tank inductor. This will minimize any interaction. Sometimes the placement of the varactors affect the tuning: anything close to the ground line can add capacitance and restrict the upper range. The coil can be compressed (lowers tuning range) or expanded (raises tuning range). The 1000 µF capacitor was added later. Without it the battery's internal resistance is often too high. Also try a pot in place of the 10k source resistor. Adjust the regeneration so that the least amount of voltage is used (5V), this enhances selectivity. Properly setup, the Heliosdyne can hear a new station with every small turn of the tuning resistor.

2.3 Dee/Mitch-dyne II
ONE-TRANSISTOR 1.5-VOLT DX
©2011

The Dee/Mitch-dyne II is a single-transistor regenerative radio. The circuit uses common-collector mode at RF and common-emitter mode at AF. Coverage is 60M to 31M; but is alterable. A MPSA18 NPN transistor (1.5 dB noise figure) provides an hfe, or current gain, of ~570. Similar; two, three, and four transistor; receivers were designed by *Sir Douglas Hall*, as far back as 1964.

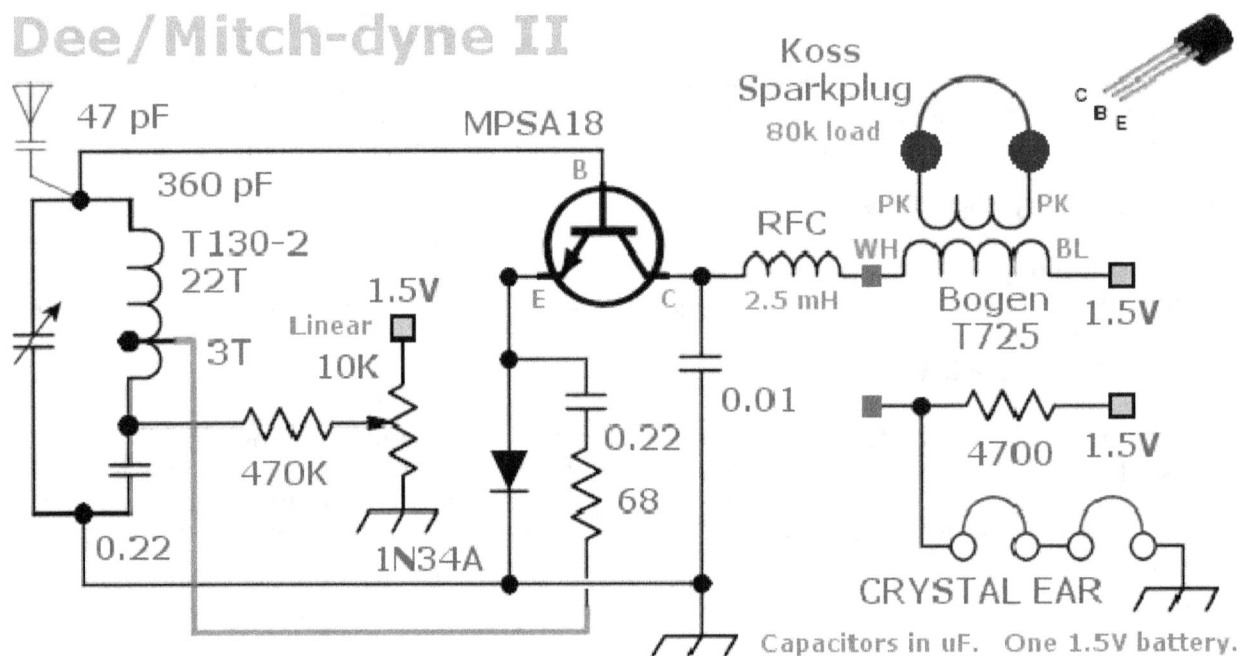

The Dee/Mitch-dyne II design boasts: only 14-parts, 1.5 Volt power, low-cost, effortlessly miniaturized, and high performance. An emitter diode *blocks half* the radio energy, which is then fed back into the tank (above using a Hartley configuration). The fixed resistor (alter to suit your needs) on the emitter creates 38k ohms of input impedance. Input impedance equals load times hfe. Regeneration is controlled via base current. The alternative design (see squares) uses two crystal earphones in series and *no AF transformer*. Using a T50-2 toroid and trimmer pots, a hobbyist can *shoehorn the radio to fit inside a matchbox*. Also, unlike a JFET, a BJT is very robust.

The Dee/Mitch-dyne II solves the problems that are usually attributed to BJT regenerators. Tank Q is maintained by a high input impedance. Yet our BJT has a higher transconductance than a FET, triode, or pentode. The fact that a BJT's current and voltage gains are not driven by its supply voltage was validated by use of a 1.5V power supply. The Dee/Mitch-dyne II is both sensitive and selective and will change how you view bipolar junction transistor regenerative receiver performance.

This circuit was named after my mother and father.

2.4 Dee/Mitch-dyne
1AD TRANSISTOR SHORTWAVE RADIO
©2008

MPSA18

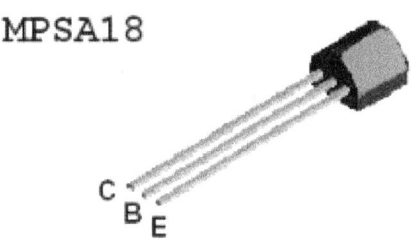

1. ABSTRACT

The Dee/Mitch-dyne is a single-transistor regen-reflex radio capable of driving earphones. The circuit uses common-collector mode for RF and common-base mode for AF. The radio covers 31M, 41M, 49M, and 60M SW bands. Only one potentiometer is necessary for regenerative control.

2. MPSA18 TRANSISTOR

My MPSA18 NPN transistor had a measured hfe of 570 or +55 dB of gain. The noise figure for the device is low. The transistor is mentioned on *Charles Wenzel*'s website (*www.techlib.com*).

3. Sir Douglas Hall

While in school I designed several circuits. After a recent extensive search I found that I was "beat to the punch" by *Sir Douglas Hall*. His circuits adorned a UK magazine; be sure to see *Geoff*'s excellent website (*www.radioconstructors.info*). The only design of mine of this type that he did not define was a purely one-transistor regenerative. I call this circuit the Dee/Mitch-dyne.

common collector   common base

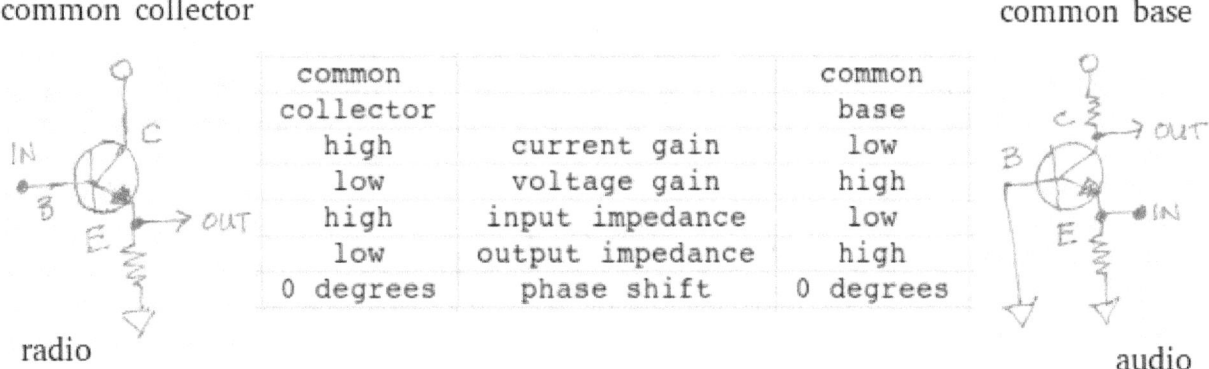

	common collector		common base
	high	current gain	low
	low	voltage gain	high
	high	input impedance	low
	low	output impedance	high
	0 degrees	phase shift	0 degrees

radio   audio

4. THEORY

The tank circuit and transformer-earphone both have high-impedance. How do we get high impedance both into and out of a transistor? Easy, we run RF from the tank into the transistor in common-collector mode, detect it at low impedance, and then output the AF through the device in common-base mode. We also siphon some of the RF energy at the detector for regeneration. It should be noted that there is no phase shift as is present in common-emitter circuits. There is first a current gain, which is perfect for driving regeneration and detection. Then a voltage gain is used to drive the earphones. My "hat is off" to *Sir Douglas Hall* for his thought provoking radio designs.

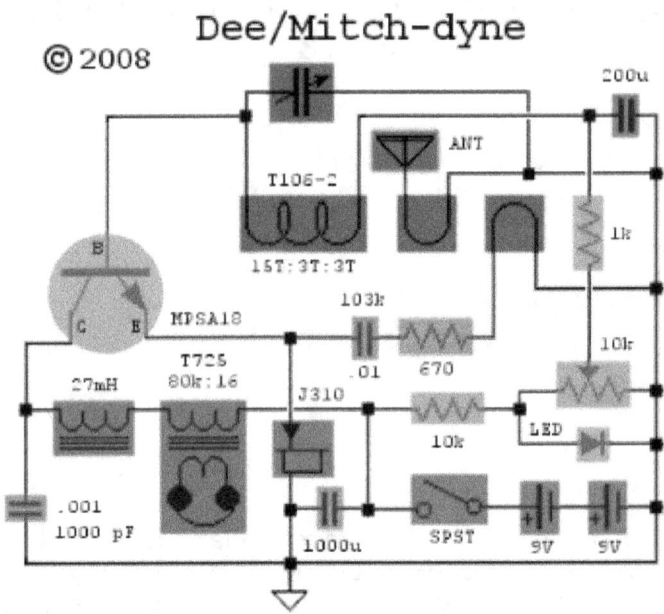

Dee/Mitch-dyne

© 2008

5. CIRCUIT DESCRIPTION

One unique feature of this radio is using a fixed regeneration circuit resistor (670 ohms). I recommend replacing this with a potentiometer for fine tuning, especially for a BCB build-up. The premise being that input impedance, for the tank, is defined as hfe times load. I ensured that the input impedance will be ~382k ohms so that tank Q will not be compromised. With a pot you can go even higher. Some "2AD" designs use this pot for regen but it often alters both Q and tuning.

Another unique feature is altering base bias to control regeneration and volume. The LED doubles as '**ON**' indicator and voltage stabilizer. The 200 uF capacitor keeps the DC headed to the base of the MPSA18. I also use a J310 n-JFET as a low impedance diode. This is accomplished by tying its drain and source together at ground. Other diodes (1N34A) worked but not quite as well. I also use low impedance 16-ohm Koss Sparkplug (thanks *Neutrodyne*) brought up to 80k-ohms via a Bogen T725 transformer. This reduces the eardrum-shattering pops experienced with piezos. Audio frequency energy passes through the tank inductor: make sure its core does not saturate.

Performance of this circuit is very good and the big stations can be heard at a comfortable listening volume. Piezo or SP phones can be used for DX but beware of noise. Sparkplug earbuds are sensitive: rated at 112 dB SPL/mW. This circuit was named after my mother, who raised three doctors and an engineer; and father, who designed over 500 devices for the military industrial complex.

Be sure to visit *Jeff Duntemann*'s 12V Space Charge page
(www.duntemann.com/12vtubes/12vtubesindex.htm).
Also see *Greg Cooney*'s Hikers One page (www.oldradios.co.nz/hikers/).

2.5 The Angelodyne Detector
Decodes AM, SSB, CW, and FM. Vacuum Tube or Solid State.
©2011

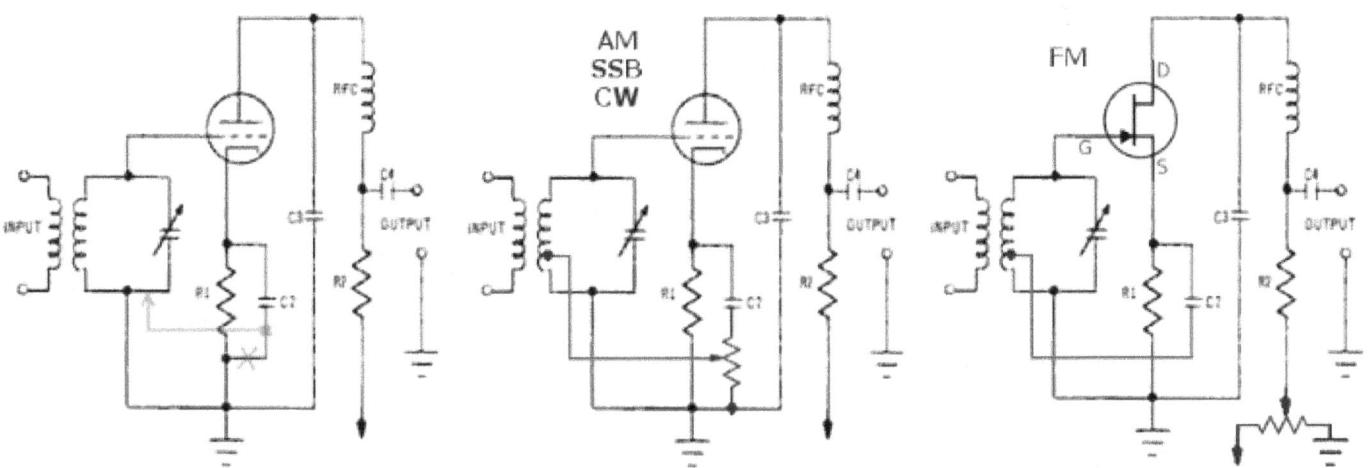

The Angelodyne detector can be setup as a *regenerative plate detector* to decode AM, CW, and SSB (Hellenedyne), or as a *regenerative drain detector* to decode N/W FM (Heliosdyne). *Dave Schmarder* noted that the Hellenedyne resembled a plate detector (see left above), except C2 injects RF energy back into the tank for regeneration (Hartley, Armstrong, Colpitts, or Clapp).

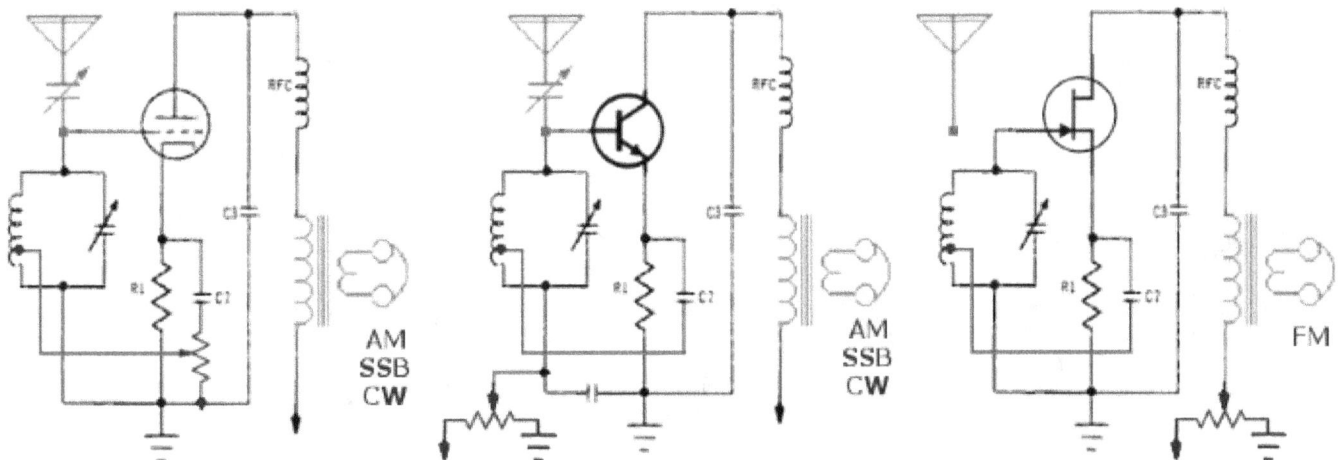

On the left the Angelodyne detector is setup as a single active device radio. It will detect AM, CW, and SSB. This was the basis for the Hellenedyne 1AD vacuum tube regenerative receiver.

In the middle the Angelodyne detector is setup as as a single active device radio. Here it detects AM, CW, and SSB. This was the basis for the Dee/Mitch-dyne 1AD BJT regenerative radio.

On the right the Angelodyne detector is setup as as a single active device radio. Here it is detecting FM. This was the basis for the Heliosdyne 1AD n-channel JFET regenerative radio.

2.6 JFET Hellenedyne
SOLID STATE REGENERATIVE "PLATE" DETECTOR
©2014

The JFET Hellenedyne is a regenerative "plate" detector that uses a solid-state J310 JFET instead of a 6GK5 tube. Regeneration is Armstrong style and controlled via a 1k-ohm potentiometer. A 10k-ohm potentiometer, set-and-forget, adjusts the operating point of the JFET. The 0.22 uF capacitor on the source leg can be replaced with one bypassing only RF: ex. 1000 pF. Select the number of tickler turns to create the smoothest regeneration. The J310 n-channel JFET has ~12 mS of transconductance. The diode prevents half of the RF from bypassing to ground.

Dave Schmarder, of _TheRadioBoard_ and _Dave's Homemade Radios_, built a MW tube version of the Hellenedyne (link here: _http://makearadio.com/tube/h-dyne.php_). He used the Hellenedyne as his _2009 Radio Contest_ and _2009 Active Device Contest_ entries. Dave went on to state: _"This has become one of my favorite dx sets as the regeneration control is very smooth."_ This is a pretty big compliment from a man who has built hundreds of radios. The tube version is showcased in the next chapter (4).

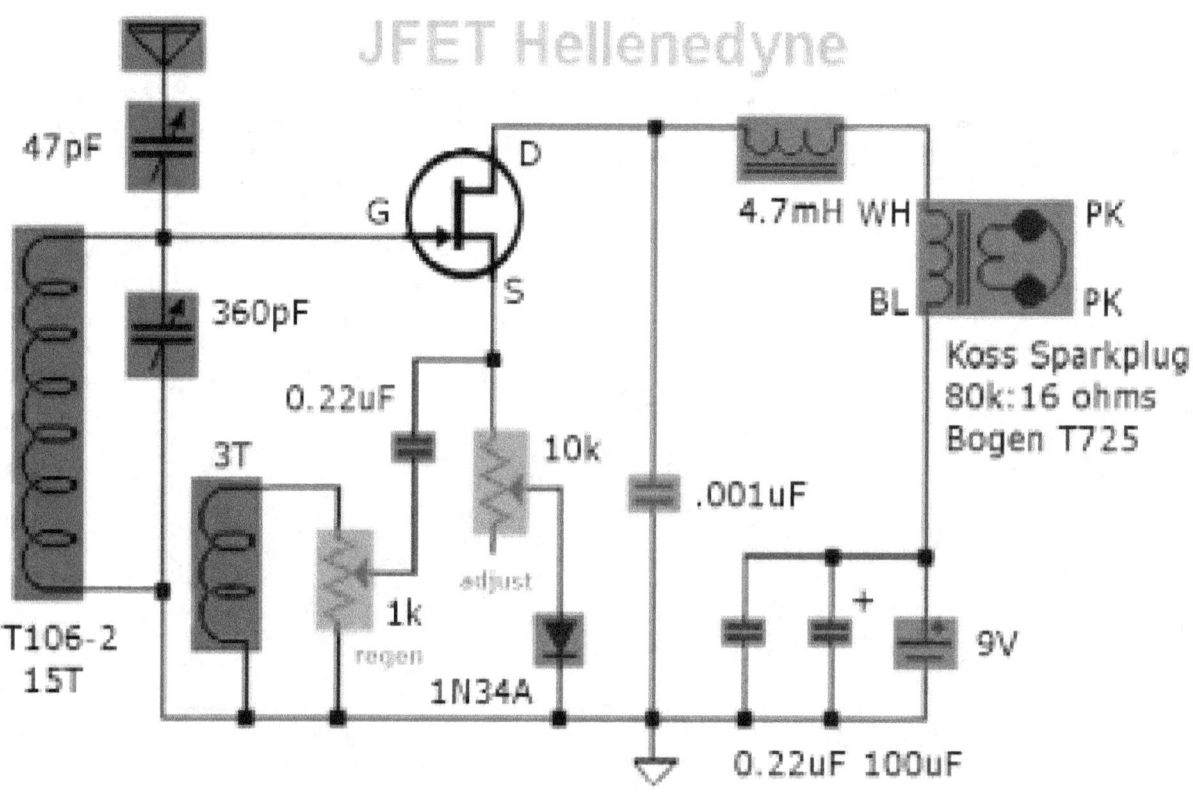

2.7 Science Fair Radio
Only eight components.
©2013

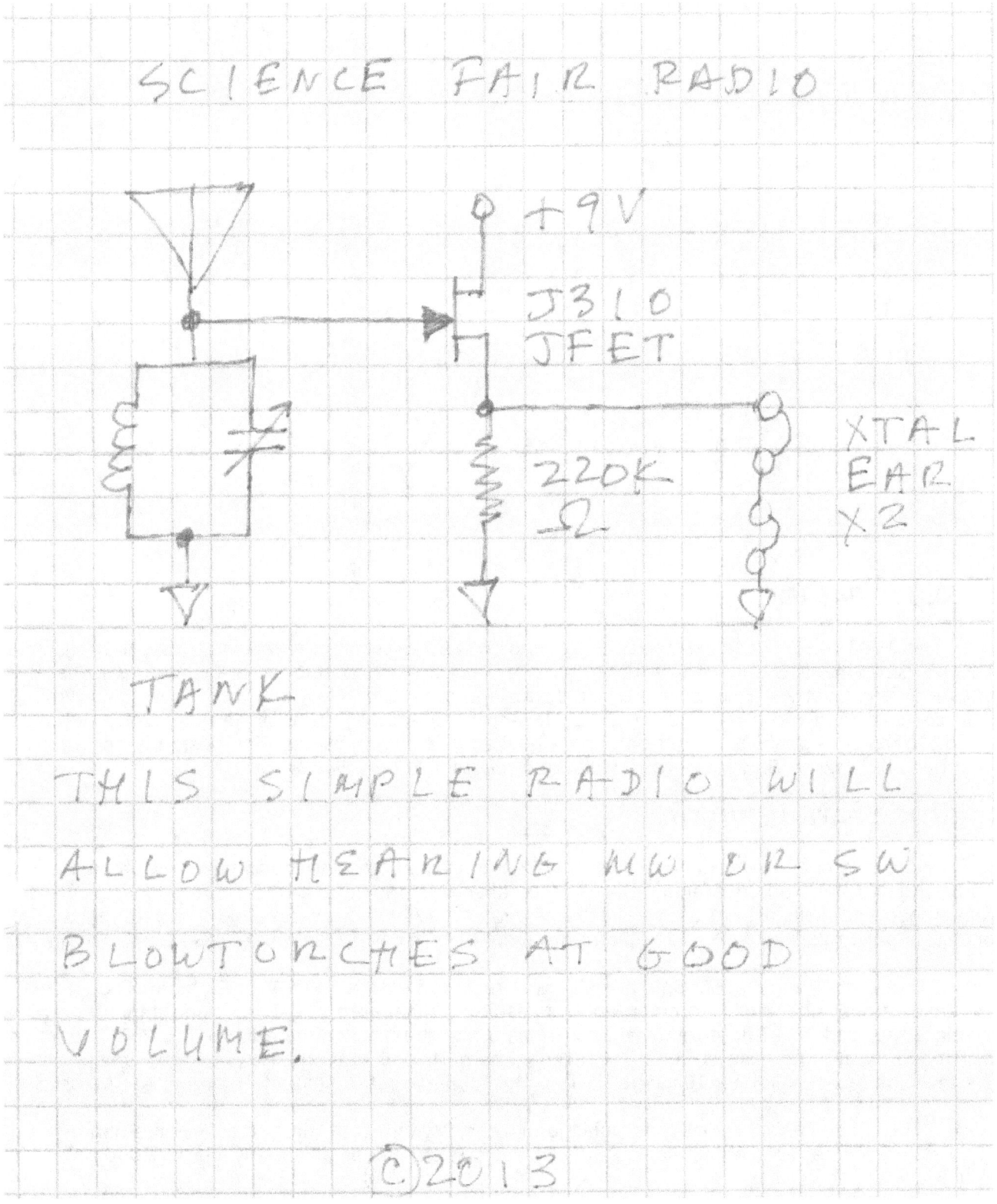

SCIENCE FAIR RADIO

+9V

J310
JFET

220K
Ω

XTAL
EAR
X2

TANK

THIS SIMPLE RADIO WILL
ALLOW HEARING MW OR SW
BLOWTORCHES AT GOOD
VOLUME.

©2013

2.8 Globe Patrol Junior
One-Transistor Medium Wave Radio
©2011

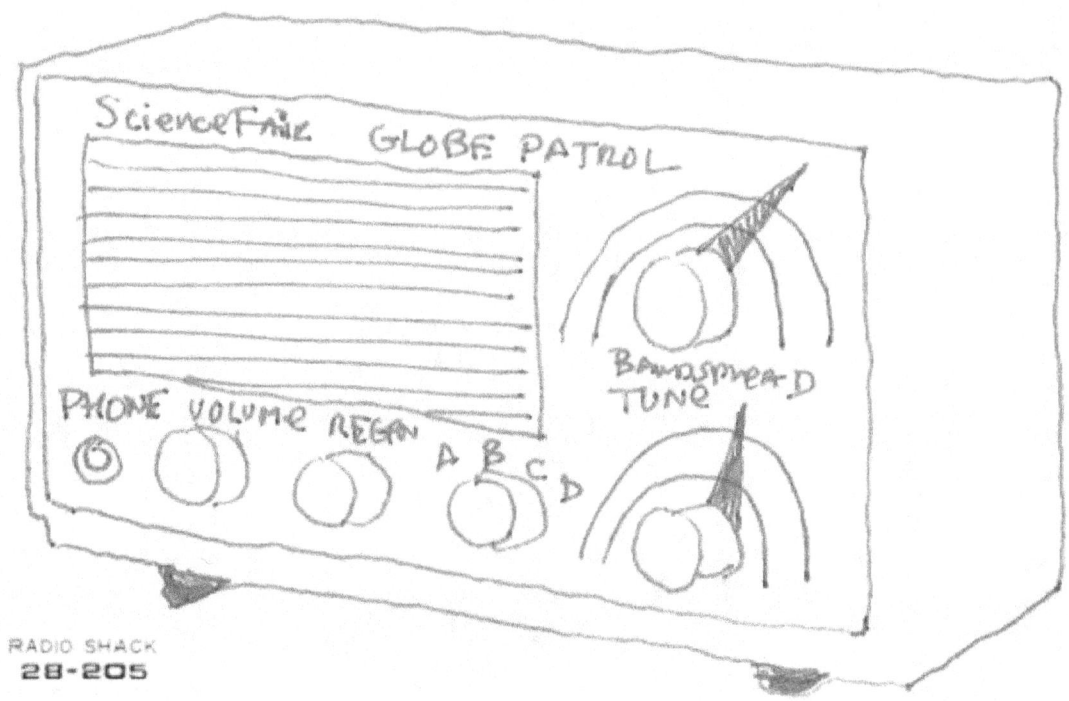

RADIO SHACK
28-205

1. GLOBE PATROL

In 1972 the *Science Fair* Globe Patrol shortwave receiver appeared in *Boy's Life* magazine. This 51-component, 3-transistor, 4-band, regenerative radio was sold, as a kit, at *Radio Shack*. It boasts a regenerative RF amplifier, diode detection, audio amplifier, and audio power amplifier. I reduced its design down to a 10-part, 1AD, low-cost, low-power, transformerless, diode detected radio for MW. The goal was to get beginners hooked on building radios. The design is not reflexed like those by: *Towler, Bazian, Wenzel, Macrohenrydyne, OldRadioBuilder, etc.*

2. GLOBE PATROL JUNIOR

Use of a crystal earphone allowed avoiding costly 2000-Ω or sound-powered phones, large audio transformers, and deafening piezos. The collector's unbypassed RFC-diode-phone circuit was researched and looks akin to a crystal set's *selectivity enhancement circuit*. The crystal ear's 15nF of capacitance takes both RF and audio to ground. There is no saturation "*pop*" like a reflexed set.

The Globe Patrol's 2SC394 transistor was replaced with an MPSA18; its high DC gain (hFE) increases input impedance. Common-emitter results in maximum power gain and its voltage gain is related to load. A BJT is more durable and has a higher transconductance than a JFET. Regen is controlled by altering base current: 3.2 µA maximum. At 600 hFE and 1.5V, current is kept below 1.9 mA. The radio, built on a 10k-Ω pot, is ideal for miniaturization: inductive tuning can be used.

The Globe Patrol's four-bands (tanks and rotary switch) were pruned to a single band. The 4 "C" batteries were replaced with one 1.5 V "D" battery: a Duracell alkaline will last ~7000 hours.

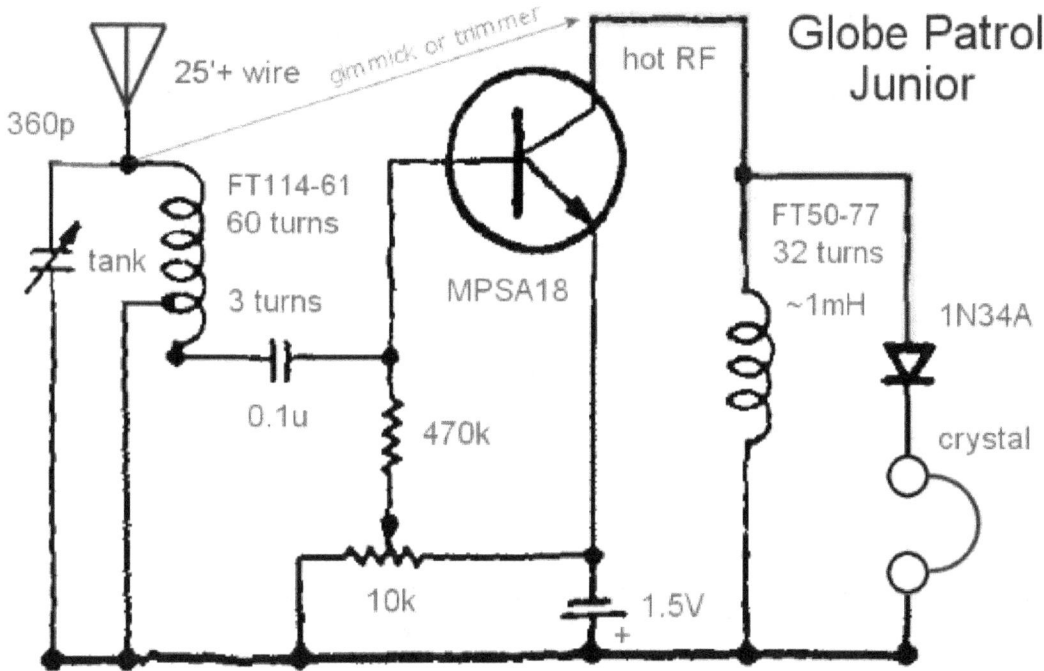

The Globe Patrol Junior uses 3-turns below ground for input into the NPN transistor's base. This provides current gain (BJT's are current controlled) and collector RF is in-phase with the tank and antenna. Collector energy provides regeneration. The 470k-Ω resistor limits MPSA18 current. Do not use over 6 volts without adding a 100-Ω emitter resistor and 0.01 µF bypass cap. With an ammeter hooked at the battery positive, rotate the pot, and mark 0.5 mA increments on the dial.

Keep all collector wiring as short as possible. The positioning of the choke is critical. Use a 1 to 10 mH choke. Crank in 1.5 mA of current, position and orient the choke, relative to the tank's inductor, until strong regeneration occurs. For more regen (a toroidal choke and tank shields RF), solder a wire at the top of the tank and lay it near the collector (or any hot RF green line above).

The Globe Patrol Junior makes a good base for experimentation. Try an antenna trim cap, antenna input taps, 720 pF tuning cap and a 43 turn coil, other diodes, a 1000 pF phone bypass, alter the 0.1 µF capacitor, use a junk radio's ferrite rod, wrap 16-turns on this rod for shortwave to ~6 MHz (or use 20-turns on a T130-2 toroid and add a trim cap from the tank to the collector), tune by moving the rod in and out of the winding, use a 100k-Ω pot (using "0" as an "off"), vary the turns into the base, etc. Or add the design to a crystal set using a few-turn base pickup coil, maximum regeneration, and physically move the coil closer to the tank. The radio is loud enough to enjoy your top local MW stations or big-gun shortwave stations. Be sure to also listen at night when transmit power is boosted. Varying base current can often be used to pull in a new station.

3. DISCUSSION

The Globe Patrol Junior captures the excitement of the Globe Patrol kit without the expense or work. It would make a good starter *one-active-device* radio for a *parent-child* team or an inexpensive science fair project.

2.9 Single Transistor Power Gain
Calculating JFET and BJT power gain.
©2014

While designing single active device radios, I became interested in the theoretical maximum power gain of transistors. A high (J310) and low (MPF102) performance JFET and a high (MPSA18) and low (2N2222A) performance BJT were evaluated using these power equations:

```
JFET POWER (common source) = gm * gm * Rload * Rin
JFET POWER (common drain)  = 0.25 * gm * Rin
JFET POWER (common gate)   = gm * Rload

BJT POWER (common emitter)   = ß * Rload / Re
BJT POWER (common collector) = ß
BJT POWER (common base)      = Rload / Re
```

Power gain in the JFET is related to transconductance, load, and input resistance. The input resistance can be from 0.1 to 1.0 GΩ. Power gain in the BJT is related to beta, load, and intrinsic emitter resistance. The intrinsic emitter resistance varies (inversely) with the collector's current.

JFET	MPF102	MPF102	J310	J310	J310	J310	J310
gm (uS)	2000	7500	8000	18000	13000	13000	13000
Rin (MΩ)	300	300	300	300	100	250	500
45k load	P (dB)	P (dB)	P (dB)	P (dB)	P (dB)	P (dB)	P (dB)
CS	155	178	179	193	178	186	192
CD	104	115	116	123	110	118	124
CG	39	51	51	58	55	55	55

BJT	2N2222A	2N2222A	MPSA18	MPSA18	MPSA18	MPSA18	MPSA18
beta	35	100	400	500	570	570	570
Ic (mA)	2.0	2.0	2.0	2.0	0.5	1.0	2.5
45k load	P (dB)	P (dB)	P (dB)	P (dB)	P (dB)	P (dB)	P (dB)
CE	102	111	123	125	114	120	128
CC	31	40	52	54	55	55	55
CB	71	71	71	71	59	65	73

Note the gain of a high-performance JFET (J310) or BJT (MPSA18). Extreme power gain is seen in the common-source mode JFET (193 dB). High power gain is seen in the common-drain mode JFET (123 db) and common-emitter mode BJT (125 dB). Reduced gains are seen in the common-gate JFET (58 dB), common-collector BJT (54 dB), and common-base BJT (71 dB). The upper horizontal bar shows the effect of input resistance on JFET gain. The lower horizontal bar shows the effect of collector current (via intrinsic emitter resistance) on BJT gain. All equations were derived by *Professor Kenneth Kuhn*.

The Angelodyne Detector (see PDF) is setup as common-drain or common-collector at radio (*for high input impedance*) and common-source or common-emitter at audio (*for high power gain*).

2.10 Realistic DX-120 Star Patrol
The nine transistor communications receiver.
©2014

1. RADIO HISTORY

One radio hobby niche is designing, building, and operating single-active device receivers. Ironically, before the advent of transistors, many tube communications receivers contained only four active devices. The devices included: a converter (mixer plus local oscillator) tube (ex. 12BE6), an intermediate frequency amplifier tube (ex. 12BA6), an audio pre-amplifier tube (ex. 12AV6), and an audio power amplifier tube (ex. 50C5). Another tube was often used in the power supply for rectification. These simple communications receivers were similar in design to the All-American Five (AA5), except with added band switching to cover both BCB and 3 or 4 shortwave bands. An added tube was sometimes used as a BFO (beat frequency oscillator) for SSB reception. And a tube for an automatic volume control circuit. Examples of this design include the: Realistic DX-75; Eddystone 870; Eico 711; Heathkit AR-2, AR3, EK-2b, GR-64, GR-91; National NC-60, SW-54h; Hallicrafters S-38, S-38E, S-41g, S-119, S-120, SW-500, WR-1000.

In the late 1960's to early 1970's, there was a transition from tubes (hollow-state) to solid state (transistors) for general coverage (non-ham) communications receivers. There was also a transition in receiver topology from single-conversion to double-conversion. During this period, some company's products disappeared with the tubes; most notably Hammarlund and RCA. Many companies went straight to solid-state, dual-conversion designs: AOR, Collins, Drake, ICOM, JRC, Kenwood, Lowe, McKay, National, Panasonic, Racal, Sony, Watkins-Johnson, and Yaesu.

But a few major manufacturers took another path: designing single-conversion, solid-state, communications receivers. These companies include: Allied (A2515), Ameco (R5), Eddystone (EC-10 Mark II, EB37, S960), Heathkit (GC1A Mohican, SW-717), Hallicrafters (S-120a, S-125), Knight Kit (R-195, Star Roamer II), Lafayette (HA-600A), Midland (11-500, 13-900), and Realistic (DX-120, DX-150, DX-150A). These radios are analog, do not cover the FM band (88 to 108 MHz), and contain no integrated circuits. Whereas, later, popular models (ex. Realistic's DX-150B, DX-160) did contain audio amplification chips. Five of the receivers above (S-120a, S-125, R-195, 11-500, 13-900) do not include a BFO knob for SSB. Note that the Heathkit and Knight Kit radios were kits. The better performers contained a sensitive "Field Effect Transistor" in their front-end: the A2515, SW-717, DX-120, and DX-150A. The A2515 was not as popular as the rest and did not include a speaker. The SW-717 was available as a kit and contains a 40673 dual-gate MOSFET. A 40673, a static sensitive device, is worth $13. Compare that to a 30 cent J310 JFET. Other random notes: the R5 was insensitive. The EC-10 was sold in the UK. The Mohican had mixed reviews; possibly due to it being a kit. And the HA-600A had mediocre image rejection.

Radio Shack's "Realistic" brand of radios was popular. Their vacuum tube DX-75 appeared in 1967. In 1968, their first solid-state general-coverage receiver hit the market: the DX-150. In 1971, the DX-150A appeared. The DX-120, introduced in 1969, was their first to include a Field Effect Transistor as the first active device (mixer). A JFET works like a tube (except with no heater): providing both sensitivity and low noise. A JFET does not load the radio frequency input tank. In 1982, Realistic introduced its "modern" line of receivers (DX-100, DX-200, DX-302).

The Allied A2515 contains a FET but tries to get away with only one transformer in the IF section, which is mated to four filters. The Eddystone 1000 is similar: it contains a FET but uses only one IF transformer and two IF filters. The modernistic 1000 can switch between two filters and also uses an audio IC. The Heathkit SW-717 uses a FET but contains no IF transformers and three filters. It also uses a Darlington pair in the IF chain. This is a compromise design. The JFET Knight Kit R-195 uses only one IF transformer and two filters. The JFET DX-120 uses an "old-school" three transformer IF section. This, basically, adds another IF transformer stage to the AA5 design. Other radios with three IF transformer designs are the: R5, EB37, EC10, 11-500, and S120a. The Mohican contains two transformers and two filters. The HA600a contains four transformers and one filter. The DX-150A, DX-150B, and DX-160 contain three IF transformers mated to one IF filter and an audio IC chip. The DX-150 and S960 contain four IF transformers.

Most radios above contain three sets of inductors: one tuning the incoming RF, one tuning the RF amplifier, and one tuning the local oscillator (which feeds the mixer). The exceptions, contain two sets of inductors, and do not have an RF amplifier before their first mixer. This design has advantages. These exceptions include the: R5, DX-120, SW-717, 11-500, and S120a.

The schematics of most of the radios above (built between 1967 and 1972) were studied. No schematics were found for the Japanese Truetone DC1270 (possible S-120a clone) or the Ameco SWL-4 (AM-only). Two radios reported in *Shortwave Receivers Past and Present* as transistor designs were found to use tubes: the EICO 711 Space Ranger and Lafayette Explor-Air Mark V. The GRE Knight Kit Star Roamer II, a transistor version of the popular Star Roamer, was determined to be similar in design to the DX-120 but with one ceramic filter and an audio IC.

Of the above, the Heathkit SW-717 and Realistic DX-120, DX-150, DX-150A, DX-150B, and DX-160 were big sellers; with the DX-150B topping the list. Of all the radios, one design stood out. The DX-120 has one of the best designs when used as a communications receiver is intended: hooked to a long wire antenna. Using a whip, it will not perform up to its potential.

2. THE DX-120

The Realistic DX-120 is a unique communications receiver, containing just nine transistors. It was designed by GRE (General Research Electronics) for Radio Shack (a Tandy Corporation). GRE also designed Radio Shack's DX-150 and DX-150A; as well as the A2515, R5, HA-600A, and Star Roamer II. GRE also designed radios for Hallicrafters and Hammarlund. Later GRE designed Radio Shack's DX-150B, DX-200, DX-300, and DX-302. The DX-120 was manufactured in Japan.

The DX-120 circuitry is as follows: four radio frequency tanks (one per band), JFET mixer, BJT (before serial No. 7211) or JFET (after serial No. 7211) local oscillator (four band inductors gang tuned via capacitor with the RF tanks), first 455 kHz IF transformer, first IF amplifier (BJT) with knob control, BFO oscillator (BJT) with pitch and on/off control, second 455 kHz IF transformer, second IF amplifier (BJT), third 455 kHz IF transformer, meter, diode detection, passive (on/off switch) automatic noise limiter, volume control (on/off switch), two BJT audio pre-amplifier, transformer with two transistor power audio amplifier, and speaker or headphone output. The power supply transformer runs two lamps and DC is conditioned via two capacitors and two zener diodes. There is also a standby switch and band spread (fine tuning) control. The dial face shows four bands: "A" 0.535 to 1.600 MHz, "B" 1.55 to 4.50 MHz, "C" 4.5 to 13.0 MHz, and "D" 13 to 30 MHz. There is a 0 to 100 logging scale and 0 to 100 band-spread scale. Controls include six knobs (off/on/volume, BFO pitch, band switching, RF gain, tuning, band spread) and three switches (BFO, ANL, standby). The front has an s-meter, headphone jack, and display. The rear has a 120V AC cord, 12V DC input, an antenna input, and switch selecting AC/DC power.

The DX-120 represents an early, popular, solid-state, analog, single-conversion communications receiver with BFO, no integrated circuits, a FET frontend, and no FM coverage. The Realistic DX-120 and Heathkit SW-717 are the two radios above with a field effect transistor front-end but no RF amplification (double instead of triple ganged tuning capacitors). This is very effective because RF amplifiers, while decreasing noise figure, increase mixer intermodulation (reduce IP3). The key is to attach these two receivers to significant antennas (50+ feet of wire). They will not work well with just a simple whip antenna. The Degen DE1103 (Kaito KA1103) is well known for its overload immunity. Degen's engineers made the DE1103 bypass its 10-dB radio frequency amplifier when an external antenna is connected. The DX-120 has no RF amplifier.

3. PL-390 COMPARISON

During my first comparison, the DX-120 using a 25 foot wire was only able to hear 70% of what the top-rated PL-390 heard. Going back and slowly tuning pushed this to 95%. Most stations were missed due to not tuning slow enough. The band-spread must be set to "100" for frequency display accuracy. Another nighttime comparison yielded 92%. This was pushed to 100% by using a trick to tune the DX-120 to exact frequencies. The trick was to listen for the local oscillator. This allows tuning exactly to the frequency of a digital receiver (a "spotter"). Tuning slowly, the next tests yielded: 96%, 100%, 100%, and 100%. Daytime DX, band D, where the DX-120 is not as sensitive, yielded 80%. I spent hours listening to hams on SSB. Something I can not do with the PL-390. It is important to note that the DX-120 would have done poorly using a small whip like the PL-390. The PL-390 contains a low-noise amplifier before its mixers: it allows solid performance using a whip. The DX-120 was simply not designed for this type of tiny antenna.

4. TRICK: LISTENING FOR THE LOCAL OSCILLATOR

The DX-120 uses high-side injection for band A or MW, band B or SW1 (1.55 to 4.50 MHz), and band C or SW2 (4.5 to 13.0 MHz). The DX-120 uses low-side injection for band D or SW4 (13 to 30 MHz). The display is off a little more and the radio is less sensitive on band D (daytime SW).

This is a description of the tuning trick. 1) Find a weak station on a spotter radio (ex. 6060). 2) On band A, B, or C set the spotter to the target frequency plus 455 kHz (ex. 6515). On band D set the spotter to the target frequency minus 455 kHz (ex. for 21 MHz use 20.545 MHz). 3) Turn down the DX-120's volume so that nothing is heard. 4) Tune the DX-120 with the intention of hearing the weak station (ex. 6060). 5) Listen on the spotter (tuned to 6515) for the background hiss to turn into silence. This is local oscillator leak. It helps to have the spotter antenna close to the DX-120. 6) Return the spotter to the original station (ex. 6060). 7) Turn up the volume of the DX-120 and compare it to the spotter. The DX-120 will be directly on station.

5. DE1103 COMPARISON

The Degen DE1103 is the reigning king of distance reception among low-priced portables. The DX-120 was compared to the DE1103, this time using the DE1103 with its three foot whip versus the DX-120 with 50 feet of wire wrap. The DX-120 was definitely the stronger radio, hearing 100% of what the DE-1103 heard; often with better results. While an unfair comparison, it shows how critical it is to have the proper antenna (50+ feet of wire) attached to the DX-120.

6. TUNING SSB

Tuning SSB is not easy with the DX-120; but possible. The BFO switch is set to "ON". Carefully use the tuning knob. Then fine tune with the band-spread knob. And, finally, use the BFO pitch knob to clean up the signal's pitch. One important **SSB tip**: reduce the RF gain so that the BFO signal (which is being injected and mixed with the audio containing frequencies) is the strongest signal in the IF chain. Make up for lost gain (less sound) by using the volume control.

Another tip: when tuning AM stations, be sure to "peak" them using the band-spread knob: shoot for the highest s-meter reading. Tune very slowly, as if expecting a station just after the current.

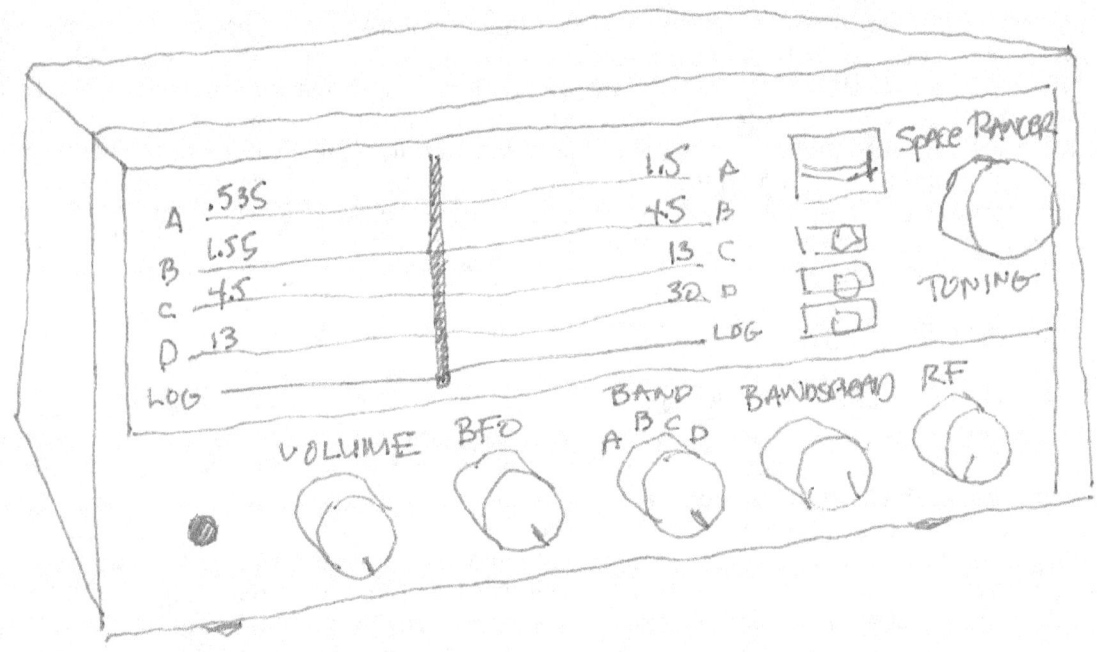

7. EICO DX-718 SPACE RANGER

I suspected that the EICO DX-718, also known as the Space Ranger, is a DX-120 clone. It has a similar analog display (only different coloring), the same three switches (only mounted vertically instead of horizontally), an s-meter (but of a different design), the same six main control knobs (only laid out differently), the same location for the headphone jack, the same left side mounted speaker, and the same rear switch, power inputs, and antenna hookup. The DX-718 is gray versus the DX-120's dark gray. The DX-718 manual (unavailable on the web) was ordered and the schematic examined. GRE never mentioned designing the DX-718 on its old website (seen using the Wayback Machine internet archive). The verdict: the DX-718 is a clone of the DX-120's "newer" version (after serial number 7211) that uses JFETs for both its mixer and local oscillator.

8. DISCUSSION

The DX-120 falls into the category of "boat anchor wannabe": it is decently built, but solid state. It is a nine transistor communications receiver. The DX-120 was the first offering from Realistic to use a FET frontend (mixer). Its high input impedance does not load the tuned RF tank. The DX-120 has a passive AGC: fading can still be heard. Band-scanning is quick and it is easy to tell which stations are strong and weak. Ironically, the DX-120 was rated higher than most of the radios with three tanks (the extra one being a tuned RF amplifier before the first mixer). Modern radios attempt to fool-proof electronics. Hobbyists wanted good whip sensitivity; so makers add an RF amplifier. This, however, reduced the IP3 of the first mixer. The solution was to run a high IP3 mixer, which are often noisy. So the radio needs to run the RF amplifier to improve noise figure. Hobbyists, in the past, got by with "AA5-like", two 455-kHz IF transformer selectivity. The DX-120 adds another IF transformer. Today, we have ceramic filters, mechanical filters, and DSP filters. The DX-120 has plenty of IF and AF gain. It lacks RF gain: the solution is to attach a 50 to 100 foot antenna, as the manual suggests. Modern digital radios can be tedious to tune through an entire band. They are stable and have memories; but can have digital noise. The new designs helped rid radios of analog displays and ganged, air-variable, tuning capacitors. But this also threw out some of a radio's mystique. Plenty can be heard on this simple set. The $70 cost of a 1969 DX-120 represents $420 in today's money. I bought mine for ~$100 on eBay, in mint condition. If you find a DX-120 for sale, you may want to snatch it up. Happy listening.

2.11 Solid State Regenerators
Commercial Regenerative Radios
©2016

Knight DX'er (3-transistor)

Science Fair Globe Patrol (3-transistor)
    NPN regenerative RF amplifier
    diode detector
    PNP audio amplifier
    PNP audio amplifier

MFJ MFJ-8100 (3-transistor, 1-IC)
    JFET RF amplifier
    dual JFET regenerative
    regen: emitter feedback
    LM386M-1 power audio amplifier IC
    RF and regen tuned by same tank

TenTec 1054 (5-transistor, 1-IC)
    varactor tuning
    JFET RF amplifier
    NPN constant current source
    dual JFET regenerative
    regen: emitter feedback
    NPN audio amplifier
    LM386M-1 power audio amplifier IC
    RF and regen tuned by same tank

TenTec 1253 (5-transistor, 3-IC)
    CD74HC4017 decade counter IC (band selector)
    varactor tuning
    JFET RF amplifier
    LM7805 voltage regulator IC
    NPN constant current source
    dual JFET regenerative
    regen: emitter feedback
    NPN audio amplifier
    TDA2611A power audio amplifier IC
    RF and regen tuned by same tank

3.1 Angelodyne
Regenerative Plate Detector
©2008

**PLATE "ANODE BEND" DETECTION
CATHODE-FOLLOWER TICKLER REGENERATION
HELLENEDYNE "High-Gm High-Rp" FRAME-GRID TRIODE
FEEDBACK-INDUCED GRID-BIAS CYCLING SUPER-REGENERATION**

1. STANDARD PLATE DETECTOR

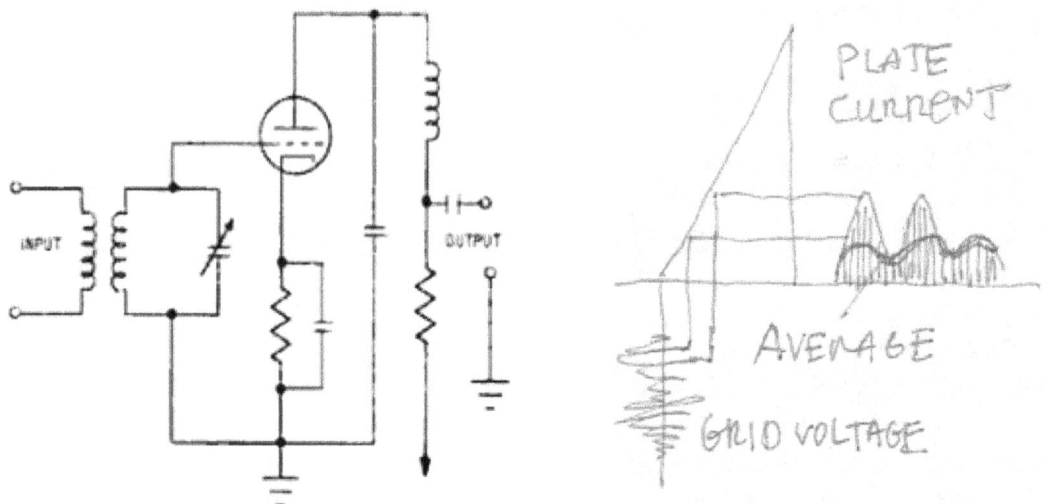

The Hellenedyne resembles a plate detector. *Dave Schmarder*, creator of TheRadioBoard, figured this out. Plate detectors are known for their **selectivity**, linearity, and input capabilities.

2. ANTENNA INPUT and HEADPHONE OUTPUT

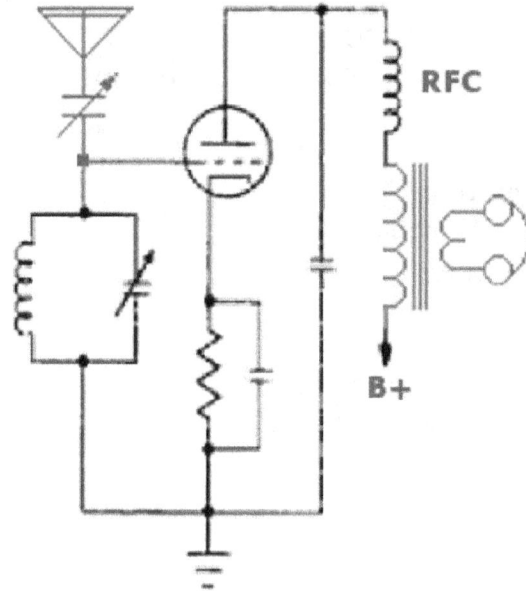

Detector input can be derived from *numerous* antenna input methods. Output is through a transformer (here *Bogen T725*) feeding headphones (here 112 dB SPL/mW 16-Ω *Koss Sparkplug*).

3. CATHODE-FOLLOWER TICKLER and HELLENEDYNE TRIODE

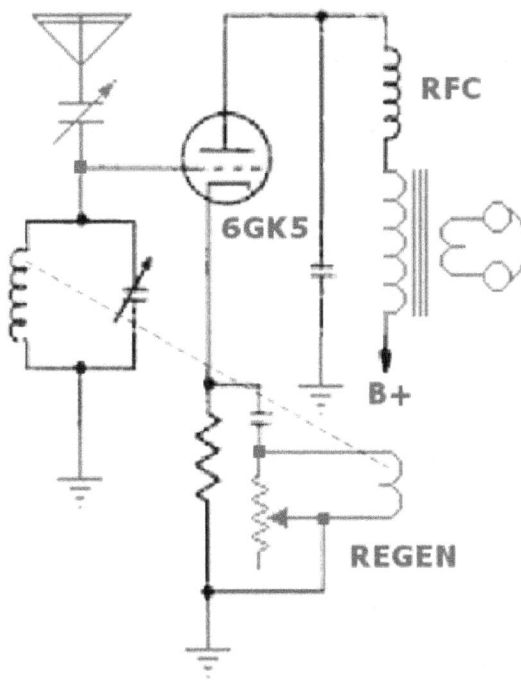

Regeneration is added via a cathode-follower (high input impedance "tank Q" protecting; non-inverted unity voltage gain; current gain only) tickler. The B7G-base 6GK5 offers both high transconductance (9000 µMHOs) and high plate resistance (5330 ohms) at 30V of plate voltage.

4. PRECISE DC GRID CONTROL

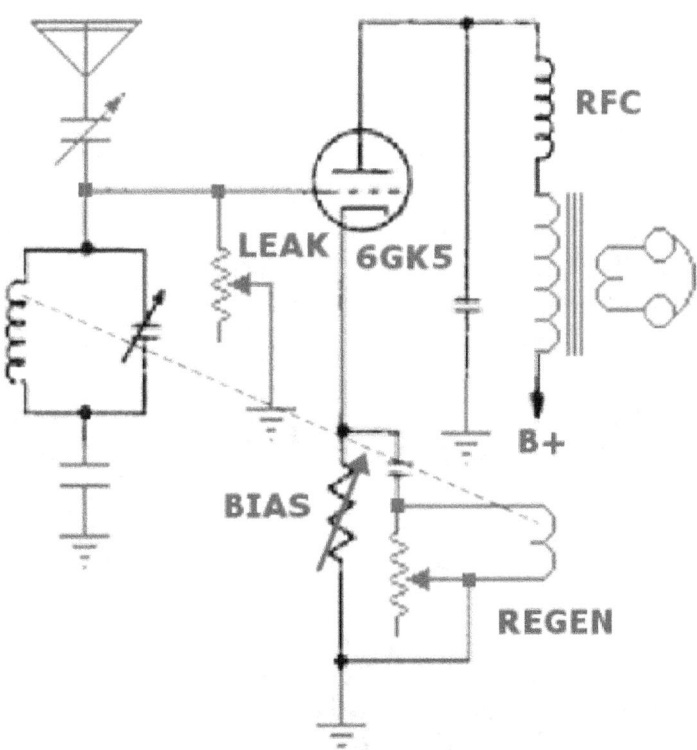

Plate detectors are biased near cut-off. Super-regeneration is doable via feedback-induced grid-bias cycling. Electrons buildup on the grid due to cathode-**bias** and are drained via grid-**leak**. Tickler electromagnetic induction stops leak: Gm falls, oscillations die, leak restarts, cycle repeats.

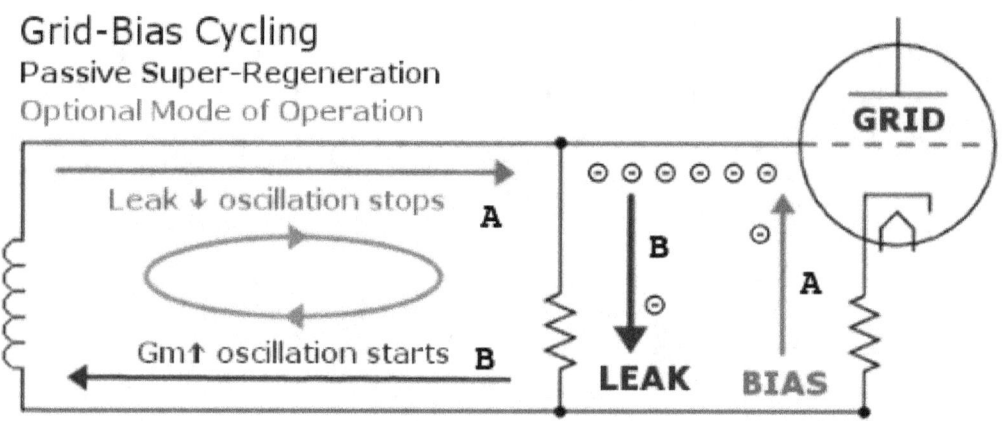

Grid-Bias Cycling
Passive Super-Regeneration
Optional Mode of Operation

Leak ↓ oscillation stops   **A**

B

Gm↑ oscillation starts   **B**

LEAK   **BIAS**   **A**

GRID

5. ANGELODYNE PLATE DETECTOR REGENERATOR

The Angelodyne is sensitive, selective, and allows for normal listening volume. My daily "shortwave listener" is currently a single-triode radio: using no lethal voltages and no sound-powered phones! Blowtorches are heard at ~80 decibels: dial-tone level.

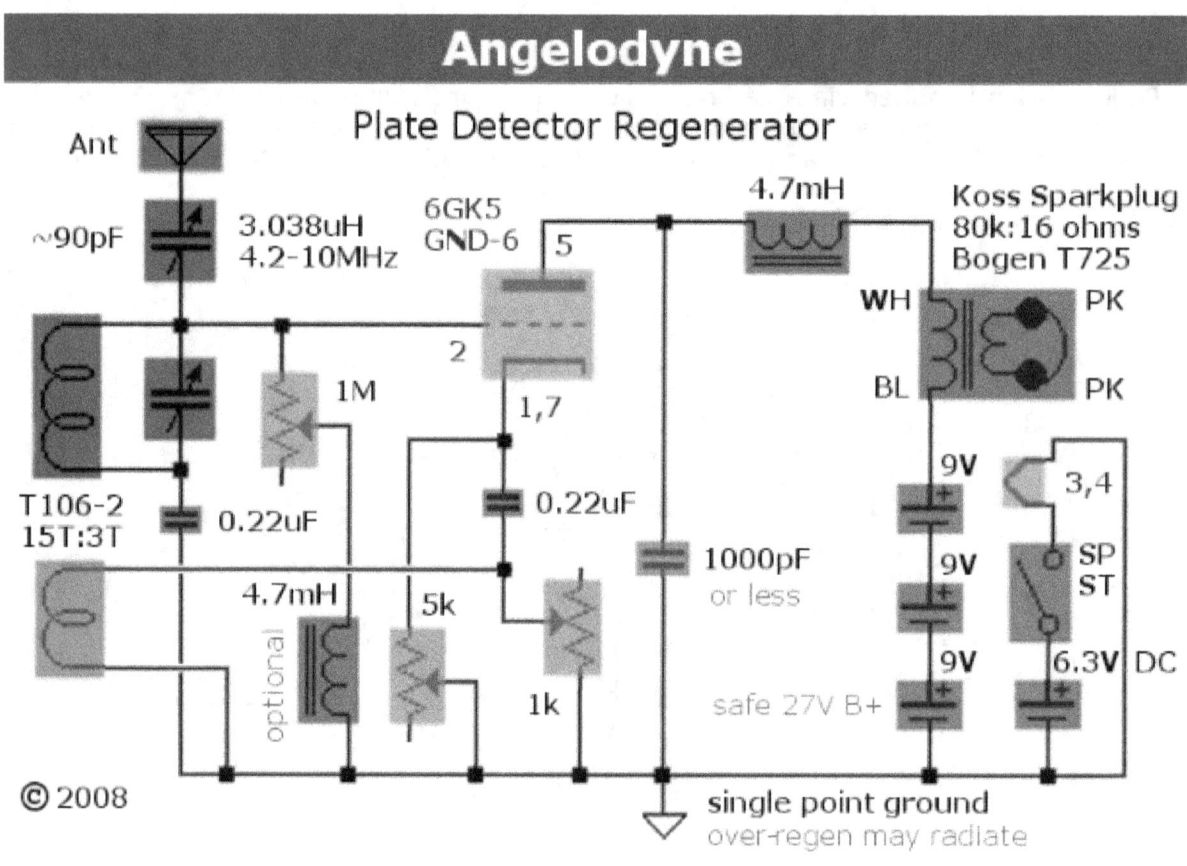

Angelodyne
Plate Detector Regenerator

Ant

~90pF   3.038uH
4.2-10MHz

6GK5
GND-6   5

4.7mH

Koss Sparkplug
80k:16 ohms
Bogen T725

WH   PK

2

BL   PK

1M

1,7

T106-2
15T:3T   0.22uF

0.22uF

9V

3,4

4.7mH   5k   1000pF
or less   9V   SP
ST

optional   1k   safe 27V B+   9V   6.3V DC

© 2008   single point ground
over-regen may radiate

I wish to thank Dave Schmarder.
http://schmarder.com/radios/tube/h-dyne.htm
http://schmarder.com/radios/
http://theradioboard.com/rb/

3.2 Hellenedyne
ONE-TRIODE REGENERATIVE PLATE DETECTOR
©2011

The Hellenedyne is a single-triode regenerative plate detector. This version uses a Hartley style of regeneration, controlled by a 10k-ohm pot. The 5k-ohm pot sets operation to near cutoff. The 0.22 µF capacitor can be replaced with one that only passes RF (1000 pF). Use whatever tank inductor tap creates a smooth regeneration. The 6GK5 is a high transconductance frame grid tube.

Dave Schmarder, who runs *TheRadioBoard* forum and *Dave's Homemade Radios* website, built a MW version of the Hellenedyne at his site: *http://makearadio.com/tube/h-dyne.php*. I was very pleased to see Dave using the Hellenedyne for both his 2009 *RadioContest* and 2009 *Active Device Contest* entries. He went on to state that: "*This has become one of my favorite dx sets as the regeneration control is very smooth.*" Be sure to visit Dave's page because his version uses a high-performance double-tuned input circuit. Dave has been building radios for over forty-five years.

The Hellenedyne is a unique type of receiver that uses both regeneration and plate detection. The detector has a high input impedance, no Miller effect, conversion gain, and smooth regeneration.

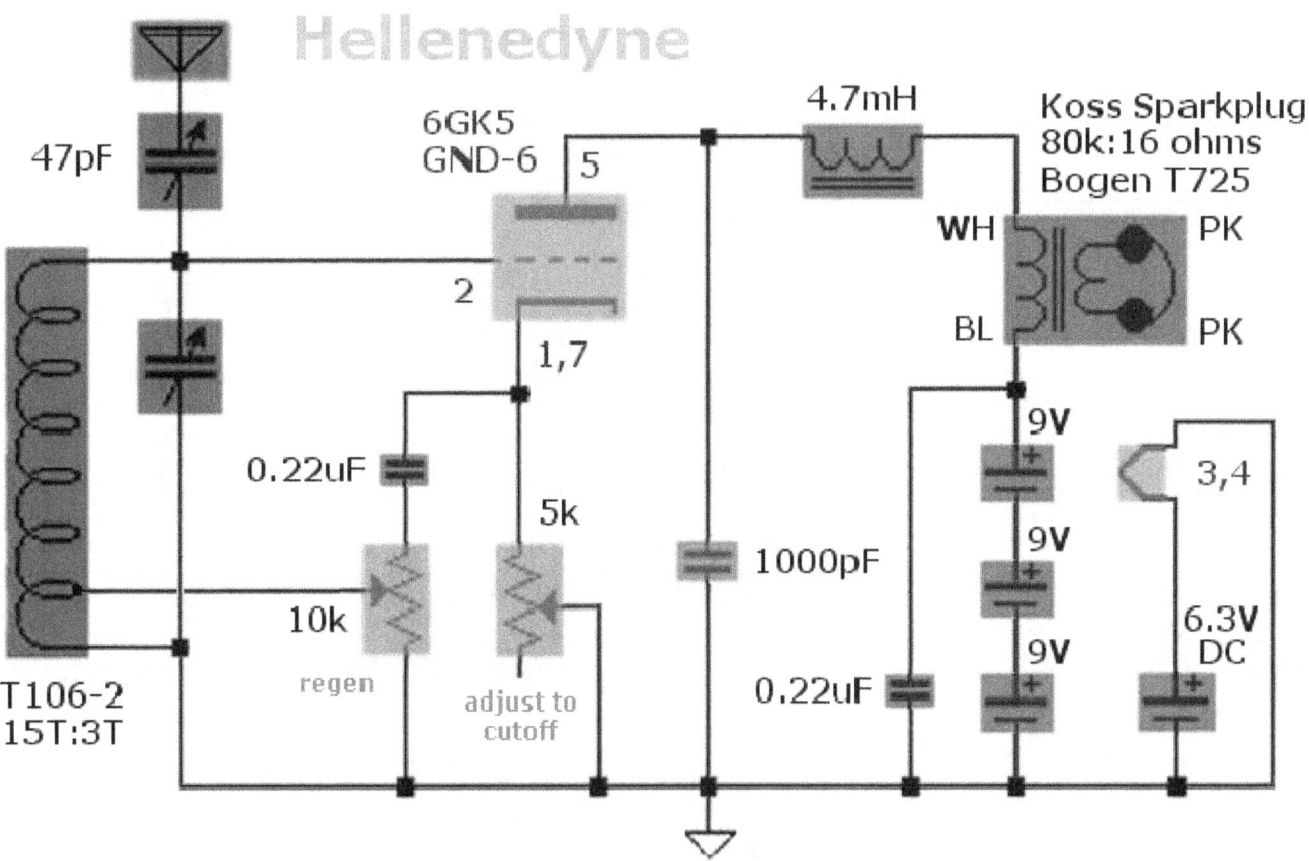

3.3 Angelodyne
MYSTERY TUBE REGENERATOR
HIGH-OUTPUT LOW-VOLTAGE SINGLE-TRIODE SHORTWAVE RADIO
©2008

May contain minor errors. See the 12-page discussion at TheRadioBoard.

1. INTRODUCTION

The Angelodyne is a single-triode radio allowing normal volume listening at reduced plate voltages. On blowtorch shortwave stations I estimate sound power levels to be near 80 decibels. This eccentric SW receiver utilizes: a Hellenedyne triode tube; cathode-follower regeneration; 16-ohm, high-fidelity, Koss Sparkplug phones (112 dB SPL per mW); and passive super-regeneration. The Angelodyne regenerator stemmed from the '*Hellenedyne radio research*', summarized below.

2. HELLENEDYNE TRIODES

Using a mathematical power model, I concluded that we enthusiasts have been using the wrong tubes in our regenerators for about 50 years. Tubes exhibiting high transconductance and high plate resistance at low plate voltages are rare. Regenerators work best at low voltages: this is common knowledge. Unfortunately at these voltages most tube's geometry allow for nearly no power output. One active device ("1AD") regenerators were forever resigned to being mated with sound-powered phones. Audio sections were often left running lethal voltages to make up for the lack of detector gain. After looking over ~14,000 valves the model identified only seven low-cost triodes. I call these Hellenedyne triodes after the circuit they were designed for. Specifically they include single (2GK5, 3GK5, 4GK5, 6GK5, 6FQ5A) and dual (4ES8, 6ES8) triode tubes. The data for the model was obtained using hand calculations from manufacturers' plate series curves. The triodes are all high-Gm, high-Rp, low-noise, low-capacitance, frame-grid VHF RF-amplifier valves.

Frame grid valves have: 1) high gain by positioning the grid closely to the cathode; 2) low noise since noise is inversely related to gain; 3) and low capacitances due to a shield between the grid and plate. They are beam triodes: technology from 1958 that competed with transistors. The B7G-base 6GK5-series valves exhibit a Gm of 7300 µMHOs and Rp of 4660 ohms at 20 V of plate. The B9A-base 6ES8-series tubes have a blistering Gm of 9060 µMHOs and Rp of 3110 ohms at 20 V of plate. These tubes are capable of excellent power output at lower plate voltages if presented with a *weak* input signal. Many tubes, especially pentodes, are anemic at low voltages after being stripped of their plate resistance. Many older high-Mu tubes achieve gain by high-Rp but have low Gm. They cannot maintain the current required to output high power, especially at lower voltages.

TUBE name	Rp ohms	Gm µMHOs	Plate mA	Plate Volts	5:1 S3 mW	5:1 S0 CW mW
7721	1980	31720	18.9	40.0	57.252	5.153
6C45	3930	12500	6.8	40.0	13.169	1.185
6ES8	2625	16000	13.9	40.0	12.714	1.144
6CW4	5410	9760	6.5	40.0	11.471	1.032
E288CC	1660	13500	21.5	40.0	3.127	0.281
12K5	840	12500	60.0	30.0	0.648	0.058
49	5200	920	5.8	40.0	0.009	0.001
6GK5	5450	11000	8.5	40.0	16.650	1.499
6GK5	5330	9000	6.5	30.0	8.747	0.787
6GK5	4660	7300	5.2	20.0	3.625	0.326
6GK5	2770	6200	3.5	10.0	0.821	0.074

Relative power output in mW (5:1 antenna coupled); S3 regen-mode or S0 CW-mode.

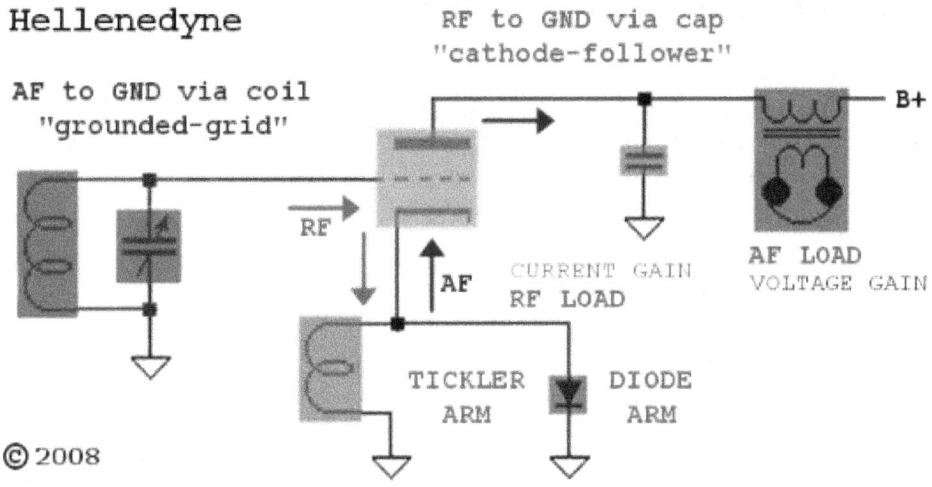

3. HELLENEDYNE CF/GG REFLEXING

The topology of the Hellenedyne was to be: regenerative cathode-follower at RF that is followed by grounded-grid amplifier at AF. This radio was *cloaked in mystery*. *Corne* concluded that the diode was not doing detection. *Dave* found that the circuit resembled a plate detector. Above we see that a tiny capacitor to ground in the plate circuit sets up cathode-follower at RF. Tank RF enters the grid and exits the cathode. The RF load consists of a tickler and diode. The tank unit to ground sets up grounded-grid at audio. Diode audio enters the cathode then exits at the plate. *Sir Douglas Hall* created an analogous transistorized circuit in 1964 called the Spontaflex.

4. CATHODE-FOLLOWER TICKLERS

Common knowledge states that High-Mu tubes make poor regenerators. This is true only with plate loaded ticklers. Cathode followers protect tank Q with their high input impedances. Voltage gain is non-inverted and reduced to unity (one) for smooth regen. Output impedance is low: perfect for diodes and ticklers! Although there is no voltage gain, current gain is significant.

5. FEEDBACK-INDUCED GRID-BIAS CYCLING

Using the Hellenedyne I found that over-saturating the toroid caused super-regeneration. Heavy tickler current was likely rendering the plate circuit inoperable in a rhythmic pattern. This was not foreign to me as I read of a similar occurrence in *J. A. Worcester, Jr.'s* 1934 Oscillodyne. The ramifications of this effect are huge. Radio Havana Cuba comes in at ~85 decibels SPL. It is likely that a minor variant will receive BCB FM and other VHF AM and FM communication. Single-triode BCB FM receiver plans are being tested. *Utilizing this effect with toroids could be new.*

It is essential to bear in mind the theory behind the Angelodyne. Audio four times greater than regeneration is possible and shortwave stations can be received with *no* antenna. Switching between regen and passive super-regen is seamless. Maybe the "sky is the limit" frequency-wise.

The theory is that tickler feedback is so great that it prevents electron flow off of the grid. **Faraday**'s Law of electromagnetic induction states that a changing magnetic field will induce an electric current and corresponding electric potential difference. Hellenedyne-tube induced heavy tickler current sets in motion a chain of events: grid potential drops, Rp/Gm drop, plate current drops, oscillations die out, negative charge leaks off the grid at a rate determined by a grid-leak resistor, and then the cycle repeats itself. *Is this part of the Hellenedyne mystery?* Larger signals disrupt the plate more often, causing less plate current, and audio is realized in the plate circuit. On MW this effect *may not* work as well due to the higher tank inductances. Critical factors may include: grid leak resistor value, toroid size, toroid turns, spacing of the toroid windings, antenna coupling capacitor value, etc. Grid bias drops because cathode electrons randomly strike the grid.

6. ANGELODYNE MYSTERY TUBE REGENERATOR

The above Angelodyne represents one of the simpler designs that worked. I believe the original Hellenedyne (*not released*) used the diode to *dampen tank Q*. The Hellenedyne can be used in a Colpitts-like mode wherein *cathode capacitance can be used for regeneration* via the internal tube capacitances! This meant no external tickler. I will be researching this more later. With the antenna removed the radio emits a high-pitched sound from the *passive super-regen*.

Mode One: low antenna coupling; maximum regeneration; increase grid-leak from zero. Mode Two: less antenna coupling going high up; grid-leak at 470k; increase regeneration from zero. Mode Three: use Two with higher grid-leak. Mode Four: use Two but over-regenerate for super-regen. Many of the components are critical to operation. Keep under 36 volts for safety.

Other aspects of the set: metal chassis, 6V DC adapter, tube upside-down, capacitors in series with main tuning capacitor for range, T725 pink wires for floating phones, 3 to 1 vernier, single point ground is floating above the chassis, chassis at potential of bottom of tuning capacitor.

7. DISCUSSION

I have been on an ambitious drive to find the '*holy grail of regenerators*'. An Angelodyne class receiver is simple, sensitive, loud, and runs off reduced voltages. The Hellenedyne did not work correctly for many so I sought an alternative. My daily "shortwave listener" is now a single-triode receiver: without lethal voltages or sound-powered phones!

3.4 Hellenedyne
SLAYING THE 1AD MONSTER
©2008

May contain minor errors. See the 12-page discussion at TheRadioBoard.

1. INTRODUCTION

The Hellenedyne is my flagship regenerative-reflex receiver. Since I was a child I wanted to design a simple tube radio that could **"hear it all"**. Would you believe there is a single active device receiver that: 1) can hear down to the noise floor, 2) employs no extra RF transformers (just the tank), 3) has regeneration that is smooth as glass, 4) uses 16-ohm earphones, 5) runs off 18 volts of plate, 6) is composed of only ten parts, and 7) uses a single triode vacuum tube?

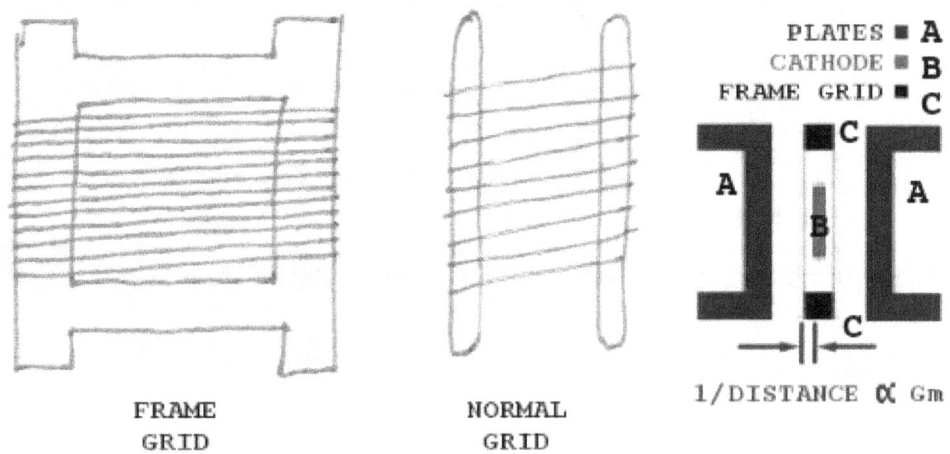

FRAME
GRID

NORMAL
GRID

PLATES ■ A
CATHODE ■ B
FRAME GRID ■ C

1/DISTANCE ∝ Gm

2. 3GK5 TRIODE

Recently I read my first 450 page vacuum tube theory book. Using the new knowledge and sophisticated power models I arrived at a unique triode, the **3GK5**. This modest **$5** B7G-base tube can be run off two "D" type and two "9V" type batteries for about 30 hours. This triode exhibits a large plate resistance and transconductance at low voltages. This was determined using data from manufacturer's curves. At 20 V I calculate the 3GK5 to have a plate current of 5.2 mA at 0 V grid. Plate resistance is 4660 ohms and transconductance a massive 7300 μMHOs. Space charge tubes, for example the 12K5 popularized by *Norm Leal*'s venerable regen boast 15000 μMHOs but reach only 15% of the 3GK5's power, in my circuit. The 3GK5 is a high-Mu (78) "gain-controlled" triode boasting dual cathode leads, 4.7 dB noise figure, 2.8 V heater (450 mA), and dissipation of 2.5 W.

The 3GK5 uses "frame grid" construction: wire one ten-thousandths of an inch in diameter is wrapped tightly around a sturdy picture-frame structure (photo above). Normally grids are self-supporting and spot-welded to supports. The rigid frame grid supports each grid wire via notches. A vacuum tube's gain is inversely related to distance from its cathode to grid, as well as cathode surface area. Frame grids allow production of consistently high gain (Gm) tubes. The tubes have low noise figures since noise is inversely related to Gm. Capacitances are lowered due to a shield existing between the grid and plate. Basically these are beam triodes. Frame grids are technology from 1958 when transistors started taking off. These rugged bottles were used for video and radar.

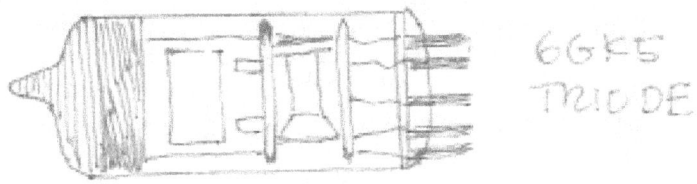

6GK5
TRIODE

3. 1AD DESIGN MUSING

How to Make a
One-tube Reflex Set
That's a "Knock-Out"

Kenneth Harkness
November 1923

Most 1AD radios fall into three groups: superheterodyne, regenerative, or *Harkness*-style regen-reflex. Some designs simply chop off an existing radio beyond the first gain stage and rely on sensitive (albeit intricate) sound-powered phones. I felt 1AD superheterodynes were uninviting: complex, low overall gain, and single "*can*" selectivity. Contrary to common belief a regenerative radio does not have infinite gain. Gain of ~78 dB can be achieved before breaking into oscillation; another 7 dB for using a regen in "CW" (beyond oscillation ala direct-conversion) mode. Regen is powerful enough to boost up an S0 (0.1 µV) signal 5:1 antenna coupled to about 3.875 mV.

Considering modern receivers shoot for +150 dB of gain; it behooved me to reflex. I was impressed with transistor designs like Tom Polk's Macrohenrydyne (*www.tompolk.com*), Robert Bazian's circuits on Charles Wenzel's website (*www.techlib.com*), and the Harkness of Tim Kilboy. What I, unfortunately, found in my own designs was just how cantankerous transistors could be: especially at cutoff and saturation. Mixing this with sensitive earphones often proved painful. I like the Harkness medium-Mu triode receiver; although, I wanted to do away with both the AF and RF transformers. Ironically the design below, after trying hundreds, is rather simple. Solutions are often that way. Could this simple circuit, that eluded us for 50 years, slay the 1AD monster?

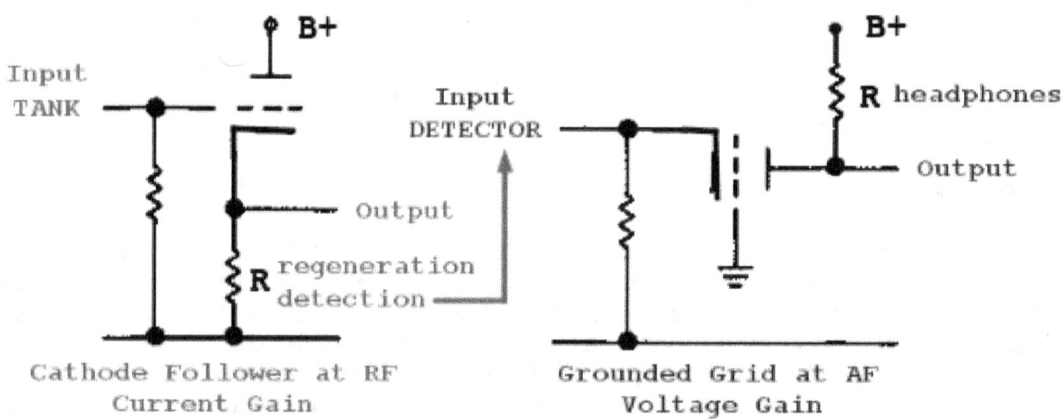

Cathode Follower at RF   Grounded Grid at AF
Current Gain   Voltage Gain

4. HELLENEDYNE CONFIGURATION

The Hellenedyne design uses the triode as a cathode-follower (common-plate) at RF and as a grounded-grid amplifier at AF. It is common knowledge amongst enthusiasts that high-Mu tubes make poor regenerative receivers. However, this is mostly true because we are using the tubes in plate loaded configuration. The primary purpose of the regenerative stage of the radio is to boost current so that the tickler and detector can be powered. Cathode followers have input impedances of 500M ohms, which protects the tank's Q. Voltage gain on the high-Mu tube is reduced to about unity (one). This buffering coupled with the non-inverting output make for a smooth regeneration. The output impedance is low, current gain is high and nearly without distortion. After rectification the audio enters the tube in grounded-grid configuration. Input impedance is low, output is high, which matches the 80k ohm earphones setup or sound-powered phones or piezos. Output is very stable, voltage gain is high, current gain is near unity, and power gain is again medium. It should be noted that many older high-Mu tubes achieve gain by having high Rp and low Gm. These tubes cannot sustain the current necessary for this setup to work properly, especially at low plate volts.

5. HELLENEDYNE CIRCUIT

The Hellenedyne circuitry above uses a 3GK5 as cathode-follower at RF and as a grounded-grid amplifier at AF. This build-up uses: a 3-turn antenna tap, T106-2 tank core (covering the 31M, 41M, 49M, and 60M bands), 25-foot antenna wire, 1000-ohm regenerative control, on/off switch, 1N34A diode, 3GK5 tube, 27 mH RF choke, Bogen T725 transformer, and 16-ohm Koss Sparkplug earphones (rated 112 dB SPL/mW). Yes, the tube is drawn inverted and pin 6 is the internal shield around the frame grid posts. The 3GK5 tube glows, runs hot, and takes ~15 seconds for warm up. Regeneration is slowly added until background "hiss" is heard. Lower pot values work even better.

There can be many circuit variations. The tank/tickler can be changed for use on BCB. On filamentary tubes utilize an RFC. The potentiometer and associated capacitor can be replaced with an air-variable capacitor; however, this might alter frequency. Other diodes can be used including thermionic ones, possibly enclosed in the same tube. The RF choke is replaceable with a "dual RC" network. The low-sensitivity 16-ohm phones can be replaced with sound-powered elements. This circuit can be run at higher plate voltages but I do not recommend anything over 45 volts as this can be dangerous and is unnecessary. Other workable tubes include: 2GK5, 4GK5, 6GK5, 6FQ5A, etc. Triode-strapped pentodes can be used but at low plate voltage they have no advantage in my circuit. The $40 US legendary 7788 at 70 V puts out only 50% the power of a 3GK5 in this design.

Performance is great: good volume, smooth regeneration, and capable of usage for DX. The receiver is cheap, simple, and easy to operate.

3.5 QRP TUBE SUPERREGENERATOR
SINGLE PENTODE SHORTWAVE RADIO
©2011

1. ABSTRACT

The regenerative radio shown above is capable of hearing, softly, the stations present on the 25M, 31M, 41M, 49M, and 60M bands. Power consumption via one pentode is a mere 17 mW.

2. CK6418 PENTODE

The Raytheon CK6418 power amplifier (2.4 mW max out) is just 1.25 inches tall. Filament power is 1.25 V at 10.0 mA; plate power is a non-lethal 18 V at 0.24 mA. Plate resistance is 420k ohms while transconductance is 300 µMHOs. The 6418 is efficient in terms of mW filament power per mA of plate current. This tube was brought to my attention by *Dave Schmarder* via his grand website. An *exhaustive* search of over 13,000 valves did not yield a finer low-power vacuum tube.

3. PHOTOS

Pictured above is the receiver as a prototype. Switch A applies power to both the filament 'A+' and plate 'B+' circuits. Potentiometer B controls regeneration. Vernier C with drive reduction (~3:1) controls the frequency. Switch D alters the antenna coupling. Jack 1 connects to the piezo headphone 2 and the terminal 3 is an antenna hookup. The 6418 is shown next to a penny below.

4. CIRCUIT DESCRIPTION UPDATED

The radio is super-regenerative. The combination of the below-tank grid-leak made up of a 470k-Ω resistor (with 1000 pF capacitor) and 9-turn tickler disrupts the grid in a rhythmic fashion. This principle is used in the Angelodyne. The RC time constant is similar to a 4.7M-Ω resistor with a 100 pF capacitor. The two arrows show the minor alterations needed to make the super-regen.

The plate circuit RF choke was replaced by two RC networks (1k-Ω resistors). This reduces radio frequency energy by about 65 dB at the cost of about 2% power loss at the piezo earphone.

The radio uses three antenna taps (3, 7, and 12 turns), a T130-2 tank toroid, Bogen T725 transformer, and 25+ feet of wire wrap for an antenna. The piezo element earphone, is similar to the *KBT-44SB-1A*. Performance is rock solid but will *never* take the prize as far as output energy. CAUTION: piezo earphones can emit a deafening pop under some circumstances (the 4.4 uF cap helps). This radio is efficient enough that it could run ~1400 hours on a set of alkaline batteries.

I wish to thank Bill "Exray" for answering many questions as I started back into the hobby. Be sure to visit Dave Schmarder's Homemade Radios website: http://makearadio.com/index.php. The **RadioBoard** is a great forum for crystal and tube radio lovers: http://theradioboard.com/rb/.

Note: I do not recommend using the KBT-44SB-1A piezo earphone. Use a crystal earphone.

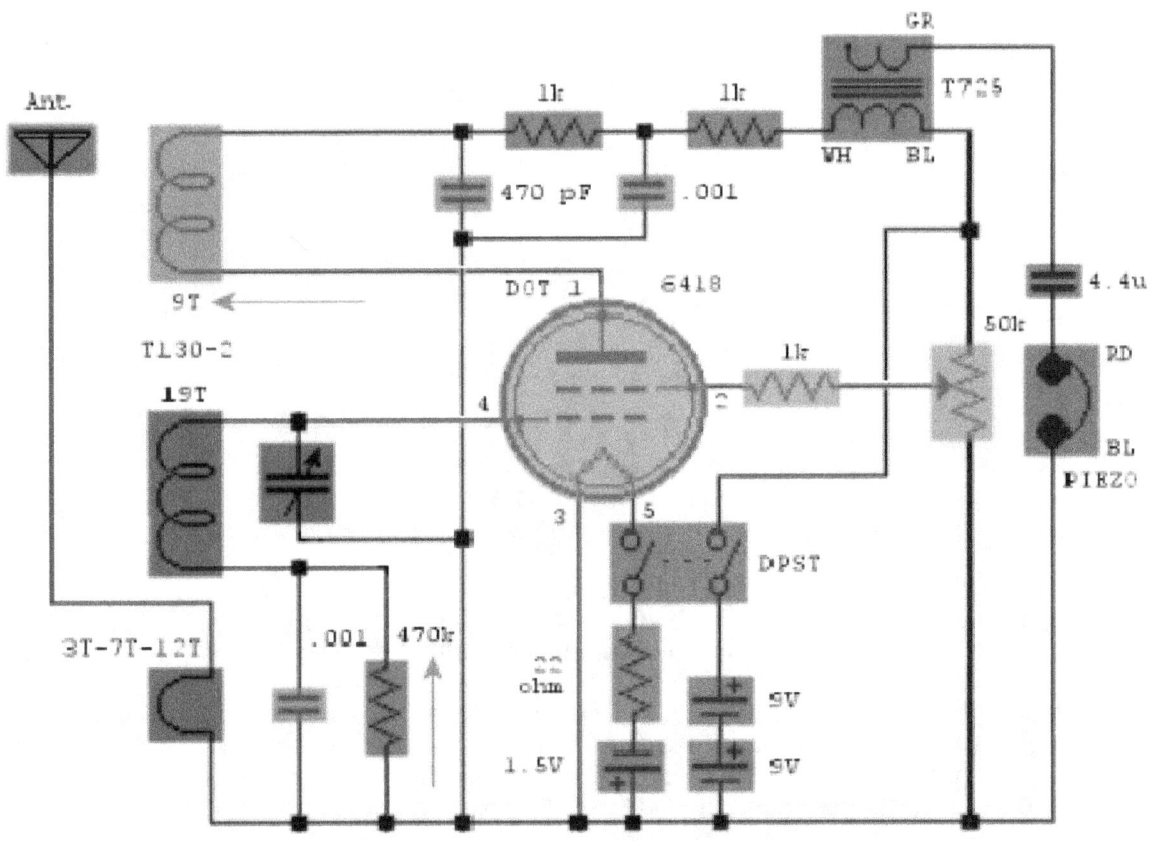

3.6 Oscillodyne
PENTODE SHORTWAVE SUPER-REGEN
©2011

1. CK6612 PENTODE

The Oscillodyne is a shortwave super-regenerative radio that uses a Raytheon CK6612 pentode. This is a 1.38 inch tall subminiature vacuum tube. Filament power is 1.25 Volts at 80 mA. It has a whopping transconductance of 3000 µMHOs; ten times that of a CK6418. Plate resistance is ~180k and plate current is 3 mA. The Oscillodyne has 40 dB more gain than a standard regenerative radio. This is because of its super-regenerative design.

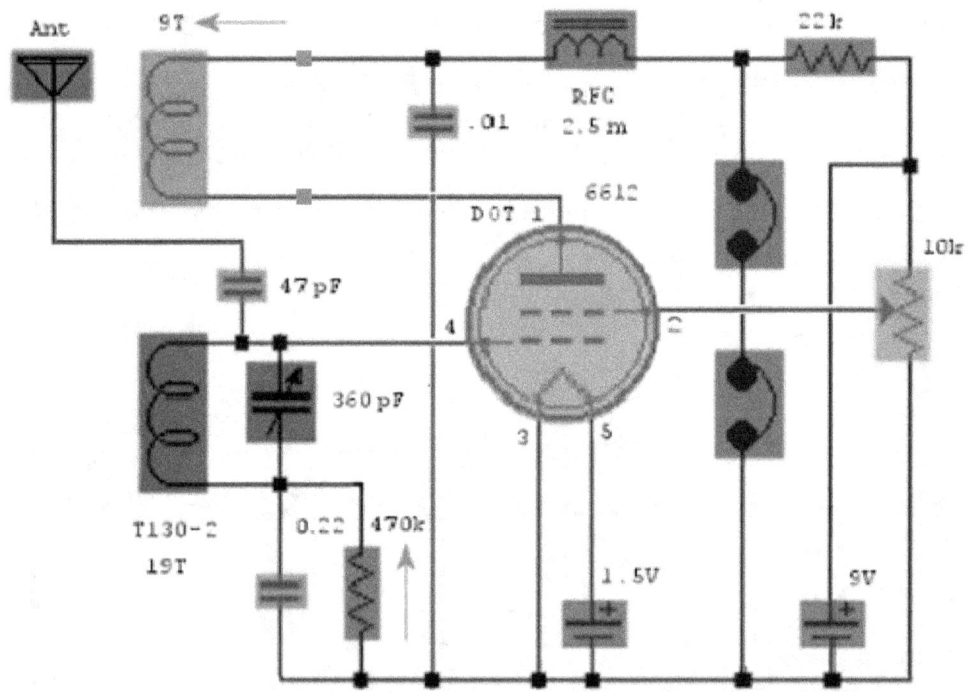

2. CIRCUIT DESIGN

The combination of the below-tank grid-leak using a 470k-Ω resistor and 9-turn tickler will disrupt the grid in a rhythmic fashion. The gain is about 120 dB or 1 million. Strong stations will sound like AM locals. Selectivity is not as good as a regenerative radio.

The Oscillodyne boasts: only 14-parts, one active device, no transformers, 9 Volts of plate, a gain comparable to some modern receivers, and a design that can be miniaturized. The radio uses two crystal earphones (in series) for immersion. A transformer (Bogen) and phones can be utilized.

The original Oscillodyne was made by J. A. Worcester, Jr. in the 1930's. I have modernized the radio by using a toroid, a pencil tube, and crystal earphones. I also altered the location of the grid-leak. The capacitor values are in µF. If the set sounds weak, simply swap the tickler lines (see the tiny squares above). Build an Oscillodyne (or convert a regen): the volume will astonish you.

3.7 Acoustic Stethoscope Headset
PIEZO POWERED
©2008

Bruce Kizerian and Dan McGillis have described using piezo elements as headphones. Dan quoted piezo sound as only ~3 dB below that of UA1614 SP elements. Recently I began attaching one Kyocera element to the chestpiece of my Tycos acoustic stethoscope for listening to my 1AD regen-reflex SW set. Might this allow more energy capture and presentation to the auditory canal?

My crude method (pictured below) uses a black heat-shrink tubing o-ring and clamps to attach the stethoscope's diaphragm closely to the piezoelectric element. The higher frequency diaphragm was used: this is made to listen to 100 to 1000 Hz heart and lung sounds. The lower frequency "bell" unit will not work as it is meant to detect murmurs in the 20 to 100 Hz range.

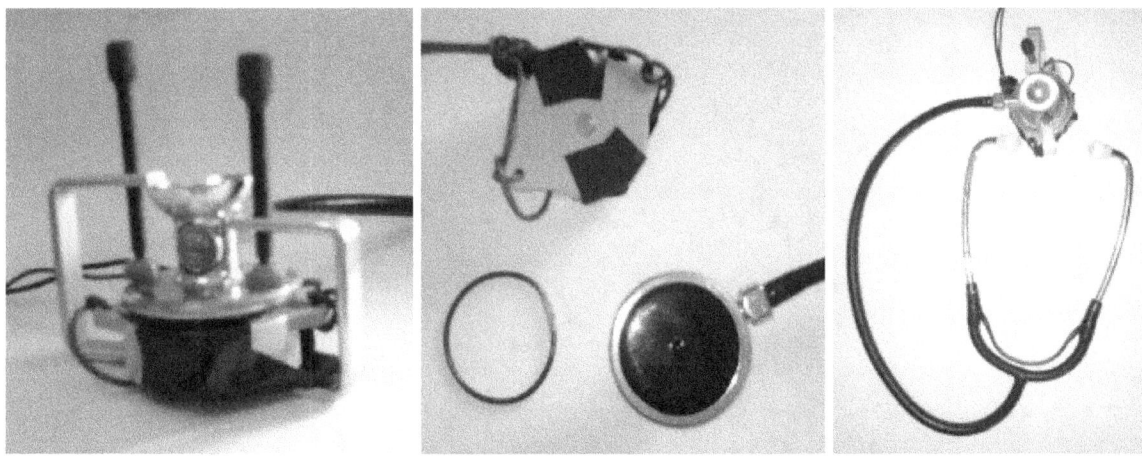

IMO All headphones share a common problem. Electricity is turned into mechanical energy; but, due to possible unintended sharp current variations, we cannot affix the element close to our eardrum without the risk of hearing damage. Consider piezo elements housed in shooter's muffs. There will be a loss of sound energy unless the elements are pressed tight against the ears. This is uncomfortable and risky. So we use low-efficiency transfer of sound through air to the ears.

By comparison the sealing ear-tips of a stethoscope are fairly comfortable. The diaphragm can be thought of as an enlarged remote eardrum that is placed close to our own via ~25 inches of tubing. There are losses at the diaphragm but the coupling at the ear canal is greater. The big plus is that unwanted energy can be filtered and dissipated by the membrane instead of the ear.

Better sound energy transfer may be possible by removing the piezo from its casing and enclosing it and the stethoscope in a cylinder. This could include baffles or deadening material to physically attenuate unwanted frequencies and protect hearing. A resultant unit might filter and shape sound, protect the ear, and transfer the high-efficiency piezo or magnet energy to the ear.

I recently hopped back into this hobby and am unfamiliar with and do not own SP phones for comparison. The stethoscope setup likely induces distortions; however, due to the diaphragm, sound is not as tinny as with the piezo elements alone. Sound is more "room filling", warmer, and bassy. My stethoscope cost over $100 but inexpensive disposable yet functional units can be purchased for ~$10. Someone with more expertise will need to decide the merits of this method.

I feel that someone might be richly rewarded by taking the time to create a proper housing and experiment with different canals, diaphragmatic materials, and piezo to diaphragm separation distance. We can replace low-efficiency "open" sound transfer to the eardrum with a low-efficiency diaphragm that can at least filter and protect the ear. If you do further testing please take care to protect your hearing. Could this close the ~3 dB gap that piezo elements have with SP phones?

3.8 Transconductance and Plate Resistance
Doing the calculations.
VERSION 1 ©2011

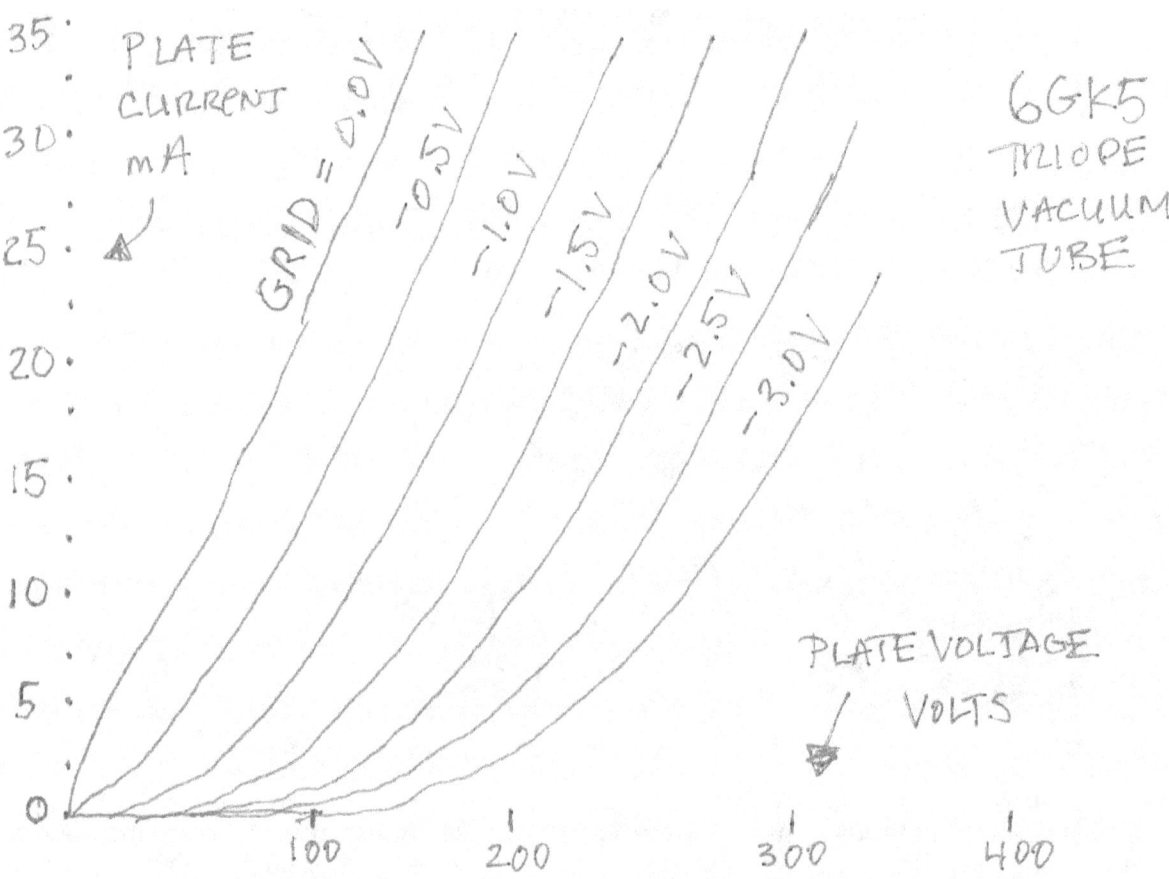

This is a quick guide to calculating transconductance and plate resistance. Values are often not given for the lower plate voltages run in our radios. Above are curves for a 6GK5 triode as per GE's data. Each line represents a plot of plate current versus plate voltage for a fixed grid voltage. This plate characteristic curve can be created: attach a 10k-Ω pot between ground and a negative supply. Then feed the wiper through a ~10k-Ω resistor and into tube's grid. Attach the cathode to ground. For a given grid voltage, alter the plate voltage and read plate current using an ammeter.

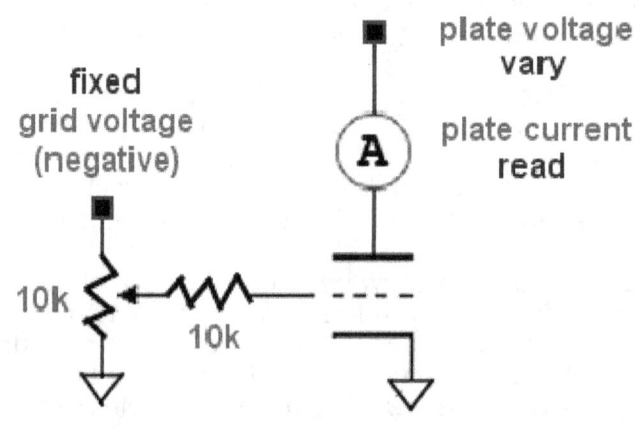

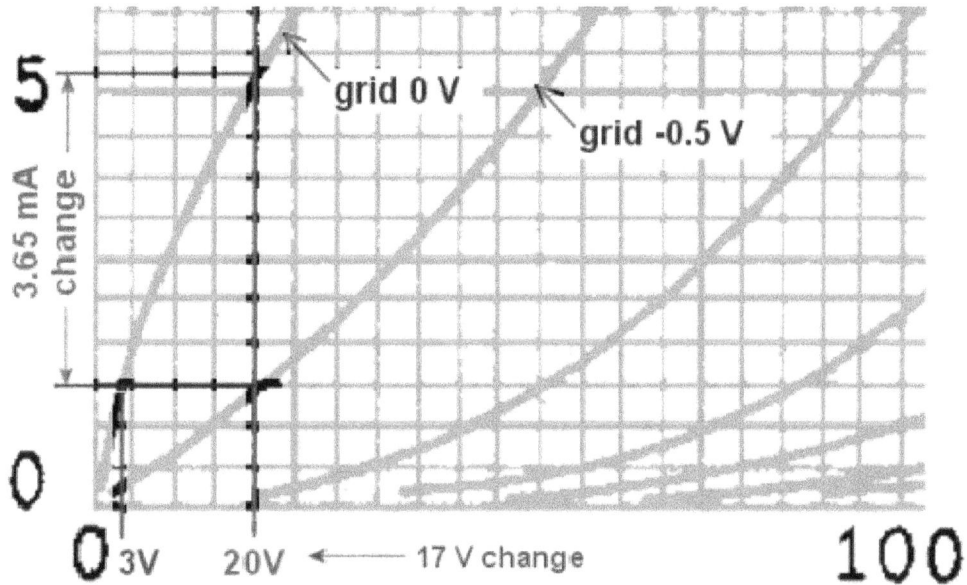

Using an editor, an actual graph was magnified. For values at 20 V, draw a line up from 20 Volts. Notice where vertical line intersects the 0 Volt and -0.5 Volt grid lines. Run each left to determine current (horizontal lines). The change between the vertical lines is 3.65 mA. Pixels may be counted using an editor and values extrapolated. Divide a 3.65 mA current change by a 0.5 V voltage change, for a transconductance of 7300 µMHOs. At 20 V, the current for 0 volts of grid is ~5.2 mA. Where the lower horizontal line intersects the 0 Volt grid line, drop a line. A change of 17 Volts (20 minus 3) is noted. Divide a 17 V voltage change by a 3.65 mA current change for 4660 ohms of plate resistance (17 divided by 0.00365). The values are then used in calculations.

$$gain = \frac{transconductance \times plate\ load\ resistance \times plate\ resistance}{1,000,000 \times (plate\ load\ resistance + plate\ resistance)}$$

For example, the values can be used in the gain equation above which appears in the *RCA Receiving Tube Manual* (*http://www.netads.com/~meo/Guitar/Tubes/t4.pl* has an online calculator). Using the Bogen T725 transformer, 16-Ω Sparkplug earbuds appear as 80k-Ω. Plugging in a 7300 µMHOs transconductance, 4660 ohm plate resistance, and 80k-Ω load, results in a gain of 30.5 or 29.7 dB (the log of 30.5 times 20). The values below were calculated, similar to the above values but for 40 Volts. The list is organized by highest gain: read from upper left, down, and then right.

Tube	plate Ω	uMHOs	mA	Tube	plate Ω	uMHOs	mA	Tube	plate Ω	uMHOs	mA
7721	1980	31720	18.9	6ES8	2625	16000	13.9	6AM4	15700	3400	2.1
7722	2120	26370	18.7	7895	5260	9500	7.3	E288CC	1660	13500	21.5
6GK5	5450	11000	8.5	6C52H	5230	8720	6.5	6DJ8	3000	9000	12.5
8627	4250	12000	12.0	6C51H	2450	13960	13.8	8058	5130	6000	4.0
6C45	3930	12500	6.8	6HA5	4625	8000	6.2	12AX7	48000	1250	1.0
6CW4	5600	10000	6.5	6HQ5	5350	7100	7.7	6H30PI	2390	6700	34.0
7462	13800	5800	3.5	6HM5	5170	7000	7.5	49	5200	920	5.8

Exceptional Tubes

D battery hours

40k_dB = gain with 40k load   ©2011 Dr. Phil

Russian	heat_volt	heat_mA	heat_mW	D_bat_hrs	plate_Ω	µMHOs	40k_dB	Construction	Description
1J24B	1.2	13	16	1,077	100000	900	28.2	lead	Pentode
1J29B	1.2	62	74	226	55000	2500	35.3	lead	Pentode

Pencil	heat_volt	heat_mA	heat_mW	D_bat_hrs	plate_Ω	µMHOs	40k_dB	Construction	Description
CK6519	1.2	10	12	1,400	300000	450	24.0	5l Sub T1-1/2x2	Pentode (F)
CK6611	1.2	20	24	700	400000	1000	31.2	5l Sub T-2x3	RF Pentode (F)
1AG5	1.2	30	36	467	180000	1000	30.3	6l Sub T-2x3	Diode / Pentode
1AG4	1.2	40	48	350	180000	1000	30.3	5l Sub T-2x3	Pentode
5672	1.2	50	60	280	20000	650	18.8	5l Sub T-2x3	Pentode (F) [power amp]
2G21	1.2	50	60	280	50000	75	4.4	7l Sub T-2x3	Triode / Heptode

Acorn	heat_volt	heat_mA	heat_mW	D_bat_hrs	plate_Ω	µMHOs	40k_dB	Construction	Description
957	1.2	50	60	233	20800	650	19.0	Acorn 5p T4-1/2	Detector Amp Oscillator

Loktal	heat_volt	heat_mA	heat_mW	D_bat_hrs	plate_Ω	µMHOs	40k_dB	Construction	Description
1LE3	1.4	50	70	200	19000	760	19.8	Loctal T-9	Medium-Mu Triode (F)

GT Tubular	heat_volt	heat_mA	heat_mW	D_bat_hrs	plate_Ω	µMHOs	40k_dB	Construction	Description
1G4GT	1.4	50	70	200	10700	825	16.9	7p Octal T-9	Triode
1G6-GT	1.4	100	140	100	45000	675	23.1	Octal T-9	Twin Triode (F)

ST Shoulder	heat_volt	heat_mA	heat_mW	D_bat_hrs	plate_Ω	µMHOs	40k_dB	Construction	Description
1B5	2.0	60	120	156	35000	575	20.6	6-prong ST-12	Duplex-Diode Triode
1H4G	2.0	60	120	156	10300	900	17.4	7p Octal ST-12	Detector Amp Triode (F)
1D7G	2.0	60	120	156	50000	400	19.0	Octal ST-12	Pentagrid Converter
1H6G	2.0	60	120	156	35000	575	20.6	Octal ST-12	Duplex-Diode - Triode (F)

Low Power Tubes

40k_dB = gain with 40k load

©2011

Tube	heat_volt	heat_mA	heat_mW	D_bat_hrs	plate_Ω	μMHOs	40k_dB	Construction	Description
CK6419	0.6	10	6	1,400	2000000	100	11.9	5-lead Sub T1-1/2x2	Pentode
6418	1.2	10	12	1,400	420000	300	20.8	5-lead Sub	Pentode (F)
CK6519	1.2	10	12	1,400	300000	450	24.0	5-lead Sub T1-1/2x2	Pentode (F)
CK6281	0.6	20	12	700	2000000	105	12.3	5-lead Sub T-2x3	Pentode (F)
1AK4	1.2	20	24	700	1500000	750	29.3	5-lead Sub T-2x3	Pentode
6088	1.2	20	24	700	700000	625	27.5	5-lead Sub T-2x3	Pentode
CK6611	1.2	20	24	700	400000	1000	31.2	5-lead Sub T-2x3	RF Pentode (F)
1AF5	1.4	25	35	560	2000000	600	27.4	7-pin Mini T5-1/2	Diode / Pentode
1U6	1.4	25	35	560	500000	300	20.9	7-pin Mini T5-1/2	Heptode (F)
1AF4	1.4	25	35	560	1800000	1050	32.3	7-pin Mini T5-1/2	Pentode
1AG5	1.2	30	36	467	180000	1000	30.3	6-lead Sub T-2x3	Diode / Pentode
2E35	1.2	30	36	467	250000	500	24.7	5-lead Sub T-2x3	Pentode [power amp]
1T6	1.2	40	48	350	400000	600	26.8	8-pin Sub T-3	Diode / Sharp-Cutoff Pentode
1V5	1.2	40	48	350	150000	750	27.5	8-lead Sub T-3	Output Pentode
1C8	1.2	40	48	350	400000	150	14.7	8-lead Sub T-3	Pentagrid Converter
1E7	1.2	40	48	350	260000	1425	33.9	8-lead Sub T-3	Pentagrid Converter
1AG4	1.2	40	48	350	180000	1000	30.3	5-lead Sub T-2x3	Pentode
1AH4	1.2	40	48	350	1500000	750	29.3	5-lead Sub T-2x3	Pentode
1AC5	1.2	40	48	350	150000	750	27.5	8-pin Sub T-3	Power Pentode
1AD5	1.2	40	48	350	700000	735	28.9	8-lead Sub T-3	Sharp-Cutoff Pentode
1W5	1.2	40	48	350	700000	735	28.9	8-lead Sub T-3	Sharp-Cutoff RF Pentode
1V6	1.2	40	48	350	1000000	200	17.7	7-lead Sub T-2x3	Triode-Pentode (F)
959	1.2	50	60	280	800000	600	27.2	Acorn 5-pin T4-1/2	Amp Triode
957	1.2	50	60	280	20800	650	19.0	Acorn 5-pin T4-1/2	Detector Amp Oscillator
2E31	1.2	50	60	280	1000000	525	26.1	5-lead Sub T-2x3	Pentode RF/IF Amp
CK5678	1.2	50	60	280	1000000	1100	32.5	Sub 1x2x3	Pentode
5672	1.2	50	60	280	20000	650	18.8	5-lead Sub T-2x3	Pentode (F) [power amp]
2G21	1.2	50	60	280	50000	75	4.4	7-lead Sub T-2x3	Triode / Heptode

Space Charge Tubes

_2 = second amplifier of tube

©2011

80k_dB = gain with 80k load

Tube	heat_volt	heat_mA	plate_Ω	µMHOs	80k_dB	mu	plate_mA	Dual-Amp Comment
12AE7	12.6	450	3150	4000	21.7	12.6	1.9	Dual Triode
12AE7_2	12.6	450	985	6500	16.0	6.4	7.5	Dual (Triode)
12AL8	12.6	550	13000	1000	21.0	13.0	0.5	Power Tetrode & (Triode)
12AL8_2	12.6	550	480	15000	17.1	7.2	40.0	Power (Tetrode) & Triode
12DW8	12.6	450		2700			1.9	Diode & Dual Triode
12DW8_2	12.6	450		6200			7.5	Diode & Dual (Triode)
12DY8	12.6	350	10000	2000	25.0	20.0	1.2	Tetrode & (Triode)
12DY8_2	12.6	350	6000	5000	28.9	30.0	14.0	(Tetrode) & Triode
12FR8	12.6	320	400000	2700	45.1	1,080.0	1.9	Diode & (Pentode) & Triode
12FR8_2	12.6	320		1200			1.0	Diode & Pentode & (Triode)
12G8	12.6							Dual Triode
12U7	12.6	150	12500	1600	24.8	20.0	1.0	+40dB Dual Triode (SAME)
6GM8	12.6	330	3400	4600	23.5	15.6	2.5	Dual Triode (SAME)

Tube	heat_volt	heat_mA	plate_Ω	µMHOs	80k_dB	mu	plate_mA	Diode-Amp Comment
12DV8	12.6	375	900	8500	17.6	7.7	9.0	2 Diode & Power Tetrode
12DE8	12.6	200	300000	1500	39.5	450.0	1.3	Diode & Pentode
12F8	12.6	150	330000	1000	36.2	330.0	1.0	2 Diode & Pentode
12DU7	12.6	250	6000	6200	30.8	37.2	12.0	2 Diode & Power Tetrode
12J8	12.6	325	6000	5500	29.7	33.0	12.0	2 Diode & Power Tetrode
12AJ6	12.6	150	45000	1200	30.8	54.0	0.8	2 Diode & Triode
12EL6	12.6	150	45000	1200	30.8	54.0	0.8	2 Diode & Triode
12DK7	12.6	500	4000	5000	25.6	20.0	6.0	2 Diode & Power Tetrode
12EM6	12.6	500	4000	5000	25.6	20.0	6.0	Diode & Power Tetrode
12FT6	12.6	150	7600	1900	22.4	14.4	2.0	2 Diode & Triode
12FM6	12.6	150	5600	2400	22.0	13.4	1.8	2 Diode & Triode
12AE6	12.6	150	15000	1000	22.0	15.0	0.8	2 Diode & Triode
12DV7	12.6	150	19000	750	21.2	14.3	0.4	2 Diode & Triode
12DL8	12.6	550	480	15000	17.1	7.2	40.0	2 Diode & Power Tetrode
12DS7	12.6	400	500	16000	18.0	8.0	35.0	2 Diode & Power Tetrode
12FK6	12.6	150	6200	1200	16.8	7.4	1.3	2 Diode & Triode

Tube	heat_volt	heat_mA	plate_Ω	µMHOs	80k_dB	mu	plate_mA	Others Comment
12CY6	12.6						?	Pentode
12DK5	12.6						?	Pentode
12EZ6	12.6	175	400000	2700	45.1	1,080.0	1.9	Pentode
12EK6	10.0	190	40000	4200	41.0	168.0	4.0	Pentode
12CN5	10.0	450	40000	3800	40.1	152.0	4.5	Pentode
12EC8	10.0	225	750000	2000	43.2	1,500.0	0.7/2.4	Pentode & Triode
12EA6	10.0	175	32000	3800	38.8	121.6	3.2	Pentode
12CX6	10.0	150	40000	3100	38.3	124.0	3.0	Pentode
12DZ6	10.0	175	25000	3800	37.2	95.0	4.5	Pentode
12AF6	10.0	150	350000	1500	39.8	525.0	0.8	Pentode
12BL6	10.0	150	500000	1350	39.4	675.0	1.4	Pentode
12AC6	10.0	150	500000	730	34.0	365.0	0.6	Pentode
8056	6.3	135	1600	8000	22.0	12.8	8.5	Nuvistor Triode
12K5	10.0	450	480	15000	17.1	7.2	40.0	Power Tetrode

Tube	heat_volt	heat_mA	plate_Ω	µMHOs	80k_dB	mu	plate_mA	Pentagrid Comment
12AG6	12.6	150		300			0.6	PENTAGRID
12EG6	12.6	150	150000	800	32.4	120.0	0.4	PENTAGRID
12FA6	12.6	150	800000	320	27.3	256.0	0.5	PENTAGRID
12FX8	12.6	270	500000	300	26.3	150.0	0.3	Triode & (HEPTODE)
12FX8_2	12.6	270		1400			1.3	(Triode) & HEPTODE
12AD6	12.6	150	400000	320	26.6	128.0	0.4	PENTAGRID
12GA6	12.6	150	1000000	140	20.3	140.0	0.3	PENTAGRID

CALCULATED PLATE RESISTANCE, TRANSCONDUCTANCE, AND PLATE CURRENT
40 VOLTS PLATE, RELATIVE GROUNDED-GRID POWER

POWER	plate	trans	mA_pl	40 Volt Calc	5:1 S3 mW
7721	1980	31720	18.9	$9 Tri-Con Calc 40V	57.252
7722	2120	26370	18.7	$20 Tri-Con Calc 40V	37.582
6GK5	5450	11000	8.5	$7 Calc 40V	16.650
8627	4250	12000	12.0	nuv $35 Calc 40V	13.522
6C45	3930	12500	6.8	$25 REAL Calc 40V	13.169
6CW4	5600	10000	6.5	nuv $17 Calc 40V	13.161
7462	13800	5800	**3.5**	NoSell Calc40V	12.987
6ES8	2625	16000	13.9	$7 Calc 40V	12.714
6CW4	5410	9760	6.5	nuv $17 Calc 40V	11.471
6CW4	5560	9460	6.5	nuv $17 Calc 40V	10.994
7895	5260	9500	7.3	nuv $16 Calc 40V	10.035
6C52H	5230	8720	6.5	nuv_R Calc 40V	7.678
6C51H	2450	13960	13.8	nuv_R Calc 40V	7.387
6HA5	4625	8000	6.2	Calc 40V	4.703
6HQ5	5350	7100	7.7	Calc 40V	4.325
6HM5	5170	7000	7.5	Calc 40V	3.887
6HM5	5400	6670	6.0	Calc 40V	3.649
6AM4	15700	3400	**2.1**	Calc 40V	3.253
E288CC	1660	13500	21.5	Calc 40V	3.127
7FC7_2	3100	9000	12.5	Calc 40V	3.120
6DJ8	3000	9000	12.5	Calc 40V	2.929
7FC7_1	3170	8500	12.0	Calc 40V	2.744
8058	5130	6000	4.0	nuv $50 Calc 40V	2.412
12AX7	48000	1250	**1.0**	Calc 40V	0.845
6H30PI	2390	6700	34.0	Calc 40V	0.778
49	5200	920	5.8	Calc 40V	0.009
5639*	2210	7200	14.5	$15 Tri-Con Calc 50V	0.829
5702*	7200	4050	5.1	Tri-Con Calc 40V	1.393
1AD4*	8080	1980	3.8	Tri-Con Calc 40V	0.201
5879*	14760	1240	2.7	Tri-Con Calc 40V	0.142
EF86*	17700	1740	2.2	Tri-Con Calc 40V	0.532

CALCULATED PLATE RESISTANCE, TRANSCONDUCTANCE, AND PLATE CURRENT
10 TO 30 VOLTS PLATE, RELATIVE GROUNDED-GRID POWER

POWER	plate	trans	mA_pl	25 to 30 Volt Calc	5:1 S3 mW
7721	2060	29050	13.7	$9 Tri-Con Calc 30V	47.510
6GK5	5330	9000	6.5	Calc 30V	8.747
PFL200*	1000	24000	47.0	$10 Pent Calc 25V	6.479
6CW4	6520	6900	4.5	nuv $17 Calc 30V	5.737
6C45	5610	7200	3.6	REAL Calc 30V	4.929
6C52H	5860	6480	4.3	nuv_R Calc 30V	3.898
5904	3225	6200	7.4	Calc 27V	1.101
12EL6	45450	1375	**1.0**	Calc 30V	1.049
12K5	840	12500	60.0	Tetrode Calc 30V	0.648
6612*	4150	4000	4.8	Tri-Con Calc 27V	0.479
6611*	10850	1240	1.4	Tri-Con Calc 30V	0.084
5904	3340	6330	7.4	Tri-Con Calc 27V	1.253

POWER	plate	trans	mA_pl	20 Volt Calc	5:1 S3 mW
7721	2145	22380	8.9	$9 Tri-Con Calc 20V	23.504
7721	2160	21880	8.4	$9 Tri-Con Calc 20V	22.264
6GK5	4660	7300	5.2	Calc 20V	3.625
7962	2090	11000	9.1	Calc 20V	2.653
6CW4	7260	4680	2.8	nuv $17 Calc 20V	2.182
7462	11100	3600	**1.8**	Calc 20V	2.130
6418*	17670	420	0.6	Tri-Con Calc 20V	0.007
6ES8	3110	9060	6.0	tt $7 Calc 20V	3.202

POWER	plate	trans	mA_pl	10 to 16 Volt Calc	5:1 S3 mW
7721	2190	15130	4.4	$9 Tri-Con Calc 10V	7.562
7721	2470	13330	4.0	$9 Tri-Con Calc 10V	6.534
6GK5	2770	6200	3.5	Calc 10V	0.821
12EL6	40400	1150	**0.8**	Calc 15V	0.527
12K5	500	15000	41.0	Tetrode Calc 16V	0.400
12EL6	36300	1100	**0.7**	Calc 12V	0.399
12K5	360	15400	34.0	Tetrode Calc 12V	0.225
12K5	370	15000	36.0	Tetrode Calc 12V	0.220

Archer Globe Patrol (3-tube)

Heathkit GR-81 (3-tube)
```
12AT7 triode regenerator
regen: variable plate voltage
12AT7 triode audio amplifier
50C5  pentode audio amplifier
35W4  diode power supply
Note: The 12AT7 is a dual triode.
```

Knight Space Spanner (3-tube)
```
12AT7 triode regenerator
regen: variable plate voltage
12AT7 triode audio amplifier
50C5  pentode audio amplifier
35W4  diode power supply
Note: The 12AT7 is a dual triode.
```

Knight Span Master (2-tube)
```
6BZ6 pentode regenerator
regen: variable screen grid voltage
6AW8A triode audio amplifier
6AW8A pentode audio amplifier
Note: The 6AW8A is a triode and pentode.
```

Knight Ocean Hopper 740 / 749 (3-tube)
```
12AT6 triode regenerator
regen: potentiometer tickler bypass
50C5  pentode audio amplifier
35W4  diode power supply
```

Lafayette Explor-Air KT-135 (3-tube)
```
12AT7 triode regenerator
regen: potentiometer tickler bypass
50C5  pentode audio amplifier
35W4  diode power supply
```

National SW-3U (3-tube)
```
1N5G pentode RF amplifier
1N5G pentode regenerator
regen: variable plate voltage
1N5G pentode AF amplifier
```